Jake Semler

Dave Thomas

Kirk Fox Jr.

Ralph Winner

Jack Sweeney

Ed Hill

Robert LeMay

Dick Simmons

Lee Kennard

Ken Wheeler

Dick Horton

Earl Woodward

Ed Hensley

For Mike & Lynda.
Happy Reading

Love
Sally

Sally T. Sweeney

Tunnel Stiffs

By S.T. Sweeney

Edited and designed by
Julie McDonald Zander of Chapters of Life
www.chaptersoflife.com
memoirs@chaptersoflife.com
360-864-6938

Printed by Gorham Printing
Centralia, Washington

Cover design by Adam Sweeney of Portland, Oregon, grandson of Jack and Sally Sweeney, performer with Adam Sweeney and the Jamboree, http://www.adamsweeney.net/

Front cover photo shows Jack Sweeney, working for Guy F. Atkinson Construction Company, high-scaling loose rock from the mountain face 300 feet above the Trinity River while working on a diversion tunnel and shaft of Monticello Dam. The photo on the back cover shows work by Gates & Fox Construction Company on the Sultan Dam diversion tunnel.

Dedication

This book is dedicated to
the hard-working miners
and their families of our era,
and the present day Stiffs
who are still changing
the face of America.

Contents

Acknowledgments

Thank you to the Guy F . Atkinson Construction Company, Gates & Fox Construction Company, the Bechtel Corporation, and John F. Bartek for their permission to use photos, newspaper articles, and other material in this book.

My sweet husband kept a detailed resume, motivating the birth of this story.

A hearty thank you to Dick Horton, Lena Thomas, Mike Gini, the late Kirk Fox, Johnnie Richardson, and others who helped jog my memory.

I appreciate my neighbors, family, and friends, who reviewed snippets of our travels as Tunnel Stiffs and found this story interesting, encouraging me to publish it in a book.

Many organizations and people provided photos used in this book, including the companies listed above, Big Earth Publications of Boulder, Colorado, Wikimedia Commons, the California Water Resources Department, National Oceanic and Atmospheric Association, and newspapers in California, Oregon, and Colorado.

Thanks to Susan Lampe and her memoir class in Tenino, Washington, for the inspiration and courage to have the story published.

A special thank you to my publisher, Julie McDonald Zander, for proofreading and editing, gently adding her assistance as needed to print and publish *Tunnel Stiffs*.

Dick Horton

"Moving from job to job, we believe that Tunnel Stiffs have made this world a better place to live by building the underground infrastructures and tunnels that provide the quality of life and the systems that we have in place today.

"Working in the mud, water, cold, heat, bad ground conditions, and all the inherent dangers of underground work, Jack Sweeney was one of the best miners in the business. I always enjoyed his humor, positive attitude, and willingness to go for whatever it took to do the job."

Dick Horton
Fredericksburg, Virginia
Fellow Tunnel Stiff

INTRODUCTION

They called us Tunnel Stiffs—nomads, always searching for greener pastures, wherever the next bend in the road took us—working, always working, building dams and digging tunnels.

This is a book about our adventures of life on the road, a road we traveled from 1949 to 1980. The main characters in this construction family are Jack Sweeney, his wife, Sally, and their son, Pat. This young family left behind their home and the coal mines of southeastern Pennsylvania for better working conditions on the West Coast.

But along the road, traveling from one job to the next, they met many other Tunnel Stiffs and their families, who became lifelong friends, more like family than blood relatives.

The story began when Corporal John "Jack" Sweeney of Plymouth, Pennsylvania, transferred from Fort Devins in Massachusetts to the Rome Army Air Force Command in Rome, New York, today known as Giffiss Air Force Base. Jack was assigned to work in the mailroom, sorting and delivering the mail, while I was a supervisor of the Mimeo, Ditto, and Embossagraph Department, my first job at the age of sixteen after growing up in Lee Center, New York.

We met in the early spring of 1944, and on June 23 of that year, Jack Sweeney and Sarah "Sally" Thomas were united in marriage at the post chapel. But our newlywed lifestyle was cut short in March 1945, when Jack received orders for overseas duty in the European Theater of Operations during World War II. He left for England, saying goodbye to a pregnant wife at what was a very, very sad time for both of us.

When the war ended in August 1945, servicemen returned home on a rotational basis. Since Jack had been overseas for only nine months, he was among the last to return home. He traveled home aboard England's *Queen Mary*, a grand ship built to carry royalty and the elite but stripped down and converted to carry troops during the war. When he arrived, Jack greeted his four-month-old son, John Patrick "Pat" Sweeney, on December 17, 1945.

After the war, a deluge of returning servicemen seeking jobs led to long lines for employment and a high demand for housing. We lived with Jack's parents in a large, three-story home in Kingston, Pennsylvania, along with our son, Pat, and Jack's two sisters and their spouses and children. We moved a few miles away to Plymouth, settling into a home with Jack's aging grandparents who welcomed us graciously. Jack returned to a job he left in 1940 when he joined the military, working long hours in the anthracite coal mines for low pay. I worked during the holiday season in a large department store in Wilkes Barre to help us purchase our discounted Christmas gifts.

After living with Jack's grandparents for two years, we found a rental house on Covert Street in Kingston, which we offered to buy, sight unseen. The owner, who, like Jack, had served in World War II, sold the three-story house to us for $200 down and $13 a month. After signing the papers, we entered the long narrow home, which was fairly new at about sixteen years old, with a small back and side yard, and fell in love with it immediately. It needed some work, but we could do it ourselves. Jack's parents and siblings helped us clean the house and move in. We couldn't afford cloth drapes, so we bought paper ones in the department stores. They looked nice and served the purpose. Jack's sister, Claire, and her husband and two children moved into the house with us, bringing furniture and appliances, while his other sister, Eileen, and her husband and daughter continued living with his parents on Landon Avenue in Kingston.

Eventually, Claire and her family moved to Clifton, New Jersey, and Jack and I finally found ourselves living as a family alone for the first time since our marriage in 1944. Jack worked ten-hour days in the coal mines, and nights in a gas station, just to pay our mortgage and provide food and other necessities. After two years, he returned home from the gas station after having worked a full day at the mines and said, "I'll be damned if I'm going to spend the rest of my life working like this just to buy a house and furniture."

I agreed with Jack, who had seen a small house trailer towed into the gas station, and it contained everything needed for housekeeping with the advantage of mobility. We both agreed we should sell our home and buy a trailer. After the house and furniture sold within a week, we packed everything we needed such as dishes, bedding, and basic housekeeping essentials in case we ever bought a trailer. We purchased a larger, used automobile with a back seat, and Jack installed a carrying rack on the roof, where he neatly stacked boxes of needed items, covered it all with a tarp, and strapped it down neatly for our trip west to California, where my brother, Tom, and his family lived. Family members in Pennsylvania grew furious when they learned we sold our beautiful home to move to California and live in a small house trailer, but it was the way we wanted to live, and a lifestyle choice we never regretted.

We stayed with my brother and his wife, Dorothy, and baby Christine while looking for a trailer, which we bought in Bellflower, California, and hauled to Tom's house. Tom and Jack dug a homemade septic tank in the back yard, and we hooked a water hose to their house and set up housekeeping.

Jack found a job with Tom's employer—the General Motors production line, which he said was required to produce one car per minute—but he found spending ten-hour days attaching the Chief Pontiac hood ornament to cars to be tiring and monotonous. Living in one place, working in one position, and meeting a quota didn't sit well with him. Jack told Tom, who was buying a home, "You might have to work this way, but I'm not going to."

Hearing of a tunnel construction job in Kernville, California, Jack hitched our trailer to the car and off we drove, embarking on a career and a lifestyle that would last three decades. With Jack working as a Tunnel Stiff, we traveled the land, building dams and drilling tunnels, changing the landscape of America forever. Pat and I joined him on this journey across the United States, where we forged friendships with unforgettable families through decades of living side by side in trailer parks, never locking our doors and welcoming newcomers to this ever-growing extended family.

Our journey continued through a blessed thirty-six years of a loving marriage.

This is the story of our lives, our love, and the lasting legacy of a generation of Tunnel Stiffs.

Tunnel Stiffs

Chapter i: Kernville, California, 1949

Isabella Dam

Our Chrysler Windsor is pulling our first trailer house en route to a job in Kernville, California.

LEAVING BEHIND THE RAT RACE and smog of Los Angeles, California, we started our life of tunnel and dam construction in 1949. Our first experience was working for Walsh Construction Company on the diversion tunnel and shaft of Isabella Dam near Kernville, California.

Jack worked as a miner. George Veatch was the supervisor and Bob McLelland oversaw the mining.

My memory is a bit vague as to the actual events.

The kitchen and dining area in our first trailer home was quite compact.

I do remember it was late summer when we arrived in Bakersfield, California, towing our very compact house—an eight-foot-wide, twenty-nine-foot-long trailer, which had a place for everything needed for our family. The bedroom had a three-quarter-size bed, with built-in drawers underneath for storage, a closet, and a dresser with a mirror on one wall. The bathroom had a small sink and chemical toilet, which I used as a hamper since trailer parks provided bathrooms and showers as well as laundry rooms and clotheslines. House trailers were just that—a place to live but easy to transport when it came time to move on.

The tiny kitchens featured a stainless steel apartment-sized refrigerator, stove, and sink, with built-in cabinets

Above is the Kernville Post Office, while at left Pat Sweeney is seen outside the town's café and bar. Below, I am pictured with a woman named Elsie on November 24, 1949, in front of the Kernville Stage Company's wagon.

along the ceiling for storage. I could stand in one spot and reach everything needed for cooking or cleaning. The table between two booth seats could be lowered and transformed into a bed if needed.

The living room had several small windows covered with Venetian blinds and drapes, an easy chair, and built-in end tables at each end of the couch, which could also be made into a double bed.

We heated with propane, with the heat rising through a floor furnace grate between the kitchen and bathroom. Ceiling vents could be cranked open to let the heat out. There were no coolers or fans at that time, either in the trailers or our cars.

We spent our first night in Bakersfield resting and preparing for the long haul up the Kern Canyon Road toward Kernville. The narrow, two-lane winding road had 200 *dangerous curves* and others not quite as dangerous Missing one of the hairpin curves meant a long plunge into the canyon below. There were no guard rails.

When we finally made it to the top, Pat and I both felt carsick and even Jack looked a little green around the gills, although he said it didn't bother him.

At Kernville, we were directed to the Rivers Edge Trailer Park, which was hot and bare of grass, like a giant cat box. The trees were gnarled cottonwoods. The few trailers in the park were haphazardly parked with plenty of space between temporary "neighbors."

We pulled in to a spot, parked, and unhitched the car from the trailer.

A quaint, old western town, Kernville was hemmed in by the High Sierra Mountains. I felt a sense of claustrophobia, especially at night. We could see only a small patch of sky showing millions of beautiful stars, but I felt like we were down in a hole.

Jack headed for the job site to see if he could hire on, while I put our house in order and started cooking supper. Pat was out the door on the run, looking over

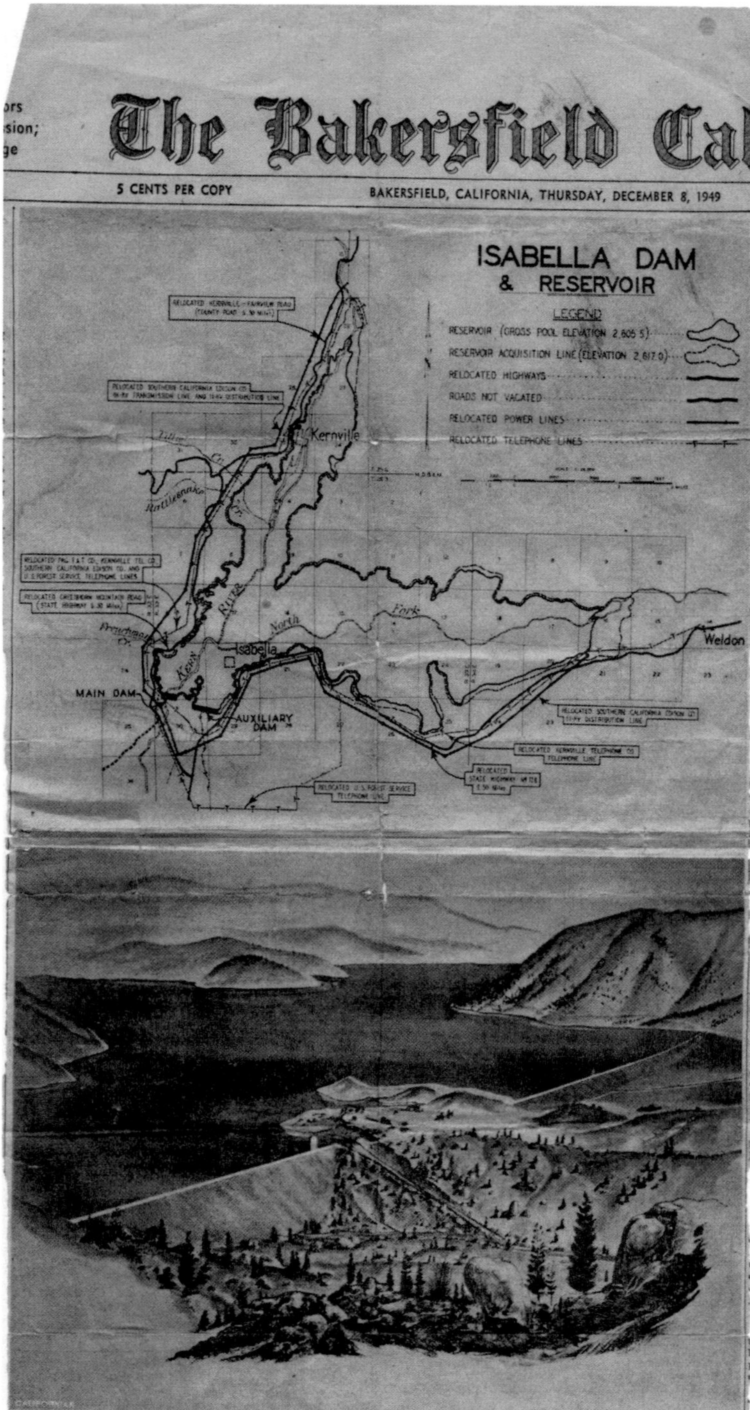

TO HARNESS RIVER—Isabella dam bulwarks begin to make changes daily in the horizon of the Kern River valley. The lower picture is an artist's sketch of the completed dam and the realigned roads and waterways. In the upper picture, the artist has shown the dam as will appear when finished. The farther structure is now finished and is an auxiliary dam using only dirt fill. Water behind this wall will never be high and according to present plans will at times be practically non-existent. The nearer wall, since it will back large volumes of water at certain times of the year, is composed of rammed earth with rock rip-rap and some concrete. The latter structure has not been started as yet.

This December 8, 1949, article and map, reprinted with permission from *The Bakersfield Californian*, shows the Isabella Dam and Reservoir that Jack helped build.

I am seen with our little dog, Poncho, in the shade of the Joshua tree in the Mohave Desert.

his new stomping grounds to see if he could find any kids to play with.

Jack told me that when he visited the job site, he introduced himself to George Veatch and inquired about finding work there. He had experience working in the anthracite coal mines in southeast Pennsylvania during his younger years. George asked if he had his "diggers" and steel-toed boots. Although he had never heard of "diggers" before, he said that he did.

George hired him on the spot, to report for work on the night shift, and sent him to meet Bob McLelland, who would be his boss on the job. Jack asked Bob, "What the hell are "diggers"? They are the tunnelers' jargon for heavy twill work pants. He stopped at the small mercantile store in Kernville on the way home and bought some Levi's.

I packed his lunch. We had supper, then rested awhile before Jack was off to his new job on the night shift.

When his shift ended, he came home with a big blister on his rear end. He had used his butt for leverage on the jackhammer he was running, resulting in a huge blister due to the rough seam in the Levi's. He said, "No more dang jeans for me," and wore his chinos (a twill fabric) from then on.

This Isabella Dam job lasted until late spring of 1950. They finished during stormy weather and the river began flooding the trailer park. We hitched up the trailer and headed for Bishop, California, where Bob and some of the other miners thought they could find work.

Jack and Pat Sweeney, above, are seen at the River's Edge Trailer Park in Kernville, California. Below left are Bob and Gerry McLelland, and below right are Mary and George Veatch in January 1950.

We headed home to Rome, New York, and found jobs with my first employer, then known as Griffiths Army Air Force Base. These photos, taken in 2011 by my niece, Rosemary Carollo, show Rome today.

Chapter ii: Rome, New York, 1950–51

Heading Home

We are riding in the car, pulling our trailer through the desert near Kernville, California, in February 1950, heading toward snowcapped mountains.

WE DROVE A SHORT WAY from Kernville to Bishop, California, where Jack rustled a job. They weren't hiring. We were a little short in the pocket for money, so we decided to pack up and head to the East Coast. Pennsylvania could always use coal miners.

We took a leisurely trip on an unfamiliar route, stopping to see places of interest along the way. In New Mexico, we visited an Indian trading post stocked mostly with goods made by the local Indians.

Going inside for a cool soft drink, we noticed the place had a dry, musty old leather smell. We saw hand-tooled leather saddles and boots, hand-loomed rugs, and so many things that were out of our price bracket.

I did buy two Zuni Indian Wedding Vases and a hammered copper necklace and earrings, which I still have. Pat got the usual little boy stuff like a tomahawk and head feathers. I don't remember what Jack ended up with except the bill.

The painting I did at left shows the Zuni Wedding Vase that we bought.

I painted a picture of my Zuni Wedding Vase in case it was broken or lost.

The Zuni Indians, a tribe of the Pueblo Native Americans, were sheepherders who lived on the prairies in little hogans made of mud and dried grass, which reminded me of Eskimo igloos.

The trading posts featured items of interest to lure travelers to stop and look. This one had a rattlesnake in a box with a glass top. We could tap on the top and watch the reptile rattle and strike. A wild animal paced back and forth in a small confined cage by the hot roadside. Several Indians in full regalia sat on the porch. Some Indian women crafted small items to sell, such as little Kachina dolls, hammered copper necklaces, and belts.

When we arrived in Kingston, Pennsylvania, Jack found a job right away in the anthracite mine where he had worked before we left.

The only trailer park we could find sat behind Wempa's Gas Station, quite a ways from town. Kingston residents didn't think much of trailer people.

I found a job in a toy factory at Kingston Corners, where I packed play gas stations all day long.

Jack's dad stayed with Pat as we returned to what seemed like the same old grind.

One weekend we drove to Upstate New York to visit my family. I stopped at Griffiths Air Force Base in Rome, where I had previously worked in the headquarters building as the supervisor of the ditto and mimeograph printing room.

They asked if I would return to work there. I told them if they could find a job for Jack, we would be happy to move back.

They hired me right away in the printing plant section of the Research and Development Division. It was two weeks before they could find a job for Jack.

Jack hitched up our trailer and we left Kingston and our Sweeney family behind, heading for New York. We parked our trailer in my sister's side yard. Jack dug a hole in the sandy ground, punched holes in a large oil drum, and added a lot of small gravel stones for a septic tank drainage field. Our "home" was set up and we were ready to work.

My sister, Mary Andrews, cared for Pat, along with her four kids, while we worked.

Jack drove a big dual-semi gasoline truck that supplied fuel for airplanes at the base. Standing only

I worked in the back of the offset area of the Research and Development Division, using the collator, which consisted of a wooden cupboard with slots for the pages. A wooden foot pedal shot the pages out in order. I stacked them into neat bundles. I also ran the mimeo, ditto, and embossaraph.

five feet, four inches tall, he had to take a running jump to leap onto the big trucks' running boards. He took a lot of kidding about that.

Our pay was virtually the same as when we left it in 1949. I was making $1.46 per hour and Jack made $1.48 an hour working ten-hour days. (Women couldn't make more money than the men.)

My job was running the mimeograph, ditto machine, collator, and embossaraph. I also made and bound the base journals. It was interesting work, but the buildings had no air conditioning at that time and it was hot! (Anyone who has lived in Upstate New York knows how hot and muggy the summers are.) We did have tall floor fans, but couldn't use them during the day as they would blow the papers everywhere.

This offset division had the first Xerox machine made, as I recall. The huge machine was built to copy and print a lot of pages and pictures in a short time, without making printing plates for the offset printing presses each time. A chemical solution had to be poured into a section of the monstrous machine so it could make copies. I thought it was a pain in the neck. This was the forerunner of our printers today. We were told, "Someday this will be our printing future."

While working here I made many friends. One of the most outstanding was John F. Bartek, a slim, balding, red-haired man in his early thirties. He was an American hero who served as a mechanic on the plane that Eddie Rickenbacker was aboard after hand-carrying a crucial message in the South Pacific during World War II. The aircraft plunged into the Pacific, leaving seven men adrift on life rafts for twenty-one days in 1942. (John, who is still alive in his nineties, remains active, giving speeches and writing stories. I found him

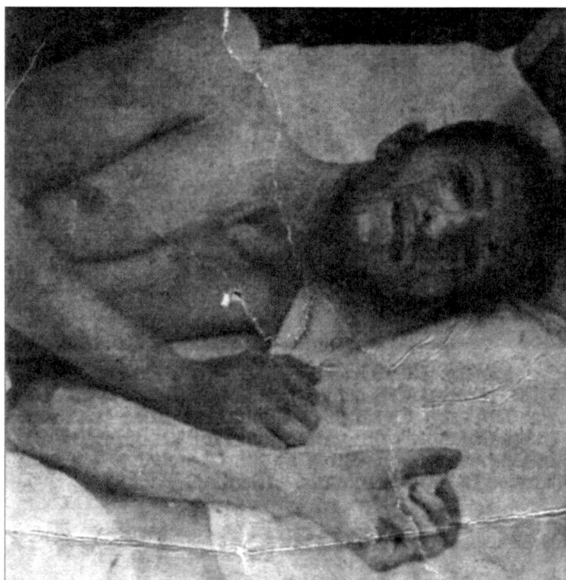

THEN—This picture in the Dec. 7, 1942 issue of "Life" showed Pvt. John F. Bartek recovering from 23 days on a life raft. He was a member of the now-famous group that crash-landed in the South Pacific with Capt. Eddie Rickenbacker.

The April 13, 1951, issue of *The Daily Sentinel* of Rome, New York, used with permission, published a photo of Private John Bartek shortly after his 1942 rescue from a life boat, where he was stranded for three weeks with Captain Eddie Rickenbacker.

Above, John Bartek, a photographer at the Griffiths Army Air Force Base, is seen on the left speaking with Frank Summa. The photo at right shows John F. Bartek and his wife, Marie, in 1998.

again a few years ago on the computer and we correspond almost daily. He gave me an autographed copy of his book, *21 Days Adrift at Sea*, which I treasure.

We worked on the base for two years, until we received news of a good job starting in Folsom, California. Tunnel Stiffs have their own construction grapevine, keeping in touch with one another and letting everyone know about new tunneling and dam construction projects. We kept in touch with George and Mary Veatch, and George sent a letter about a "good job starting up near Folsom where his good friend, Marion "Jake" Semler, was job superintendent. He asked if we would we like to see about a job there. Thinking about it a very short time, we got "itchy feet" for California and our next tunneling adventure.

The couple of years we worked on Giffiths Air Base, we managed to save enough to buy another car and a new trailer. This lovely new home was slightly larger at thirty-two feet long with a separate bedroom for Pat, complete with bunk beds, a dresser, and a closet. We had a full bathroom with a shower and toilet, a larger kitchen with counters and bigger appliances, and a combined dining and living room with a comfortable couch and overstuffed chairs. We could fold and put away the dining room table and chairs.

We purchased a used green Chrysler and maroon and white trailer with money saved working at Griffiths Army Air Force Base.

Here we go again! We packed up our new trailer and left Rome, New York, headed for a new adventure. If my memory serves me right, we took Route 66 across the country.

Our friends from Research and Development threw us a going-away party. Bob Horrigan, the head honcho where I worked, gave me his brother's address near the Grand Canyon and said to stop and see him if we were heading that way. His brother was the head of Indian Affairs in that area.

This was an easier trip, as we knew where we were going and what to expect along the way. We were on the road on Pat's seventh birthday. Jack pulled off into a trailer park along the way, knocked on doors, rustled up some kids, and held a birthday party for him. Pat was thrilled! He thought we had forgotten. He received his usual gift of cowboy boots, toy guns, and holster.

We visited the Grand Canyon in Arizona. What a wonder that was! Awesome! We couldn't begin to imagine an earthquake or flooding so fierce to create such a huge chasm.

Pat Sweeney poses before the Arizona welcome sign in August 1952. We took our leave of the Horrigans after the Grand Canyon tour and headed west for California and the Folsom Dam project.

We met Bob's brother and his wife, who invited us to spend the weekend with them and they would show us around. We ate dinner that evening in a huge restaurant and lounge on the very rim of the canyon. Glass windows enclosed the dining area. A huge thunder and lightning storm erupted while we were dining. The booms of thunder and crackle of lightning was frightening. I was sure one of those lightning bolts had my name on it.

You had to sign up in advance to take a burro trip down the winding, narrow roads to the Colorado River below. I think that trip was not for the faint of heart—definitely not for us.

The Horrigans did take us on tours around the canyon, including one down the winding roads of Oak Creek Canyon. It was a long way down, and a thrilling, bumpy ride by Jeep to the canyon floor.

The scenery looked right out of the Old West, where Hollywood directors filmed many of the cowboy movies. Some areas had small western town facades where movies had been shot. Pat really soaked that up. He was in his cowboy stage, telling everyone he was "gonna" be a cowboy when he "growed up."

We resumed our long, hot cross-country trip. Automobiles still did not have air conditioning. I kept a large bucket filled with cakes of ice. (We stopped at gas stations for ice when we filled the car with gas). I put towels on the ice and when they grew wet, hung them in the open windows. That helped a lot. Gasoline cost a few cents a gallon. Food was cheap. It didn't cost much for this trip, as we would just pull off the road to sleep and eat.

When we reached the Mojave Desert, we saw warning signs: "Water bags and gasoline" and "Last gas station for so many miles." We would drive a few more miles and would see the same signs. As a precaution, in case we became stuck in that vast desert without any water, we did buy large canvas bags of water, which we hung on the car's front bumper to keep cold. We were so naive then, bumpkins just off the turnip truck.

We kept looking for miles to see the "desert," which I envisioned as having sand dunes and blowing sandstorms. Finally, we asked at one gas station and were informed that we had been crossing it for the past few days. It was all green, as it was early spring. Traveling the desert at night was wonderful. In the heavy nighttime air, the desert blossoms gave off a beautiful and unforgettable aromatic scent.

Our cross-country trip was fantastic and unforgettable, especially our visit to the Grand Canyon. We have been back a few times, but nothing compares to that first time we saw the Grand Canyon.

This photo shows the magnificent Grand Canyon behind me.

Chapter iii: Folsom, California, 1952

Folsom Dam

Jack was an excavator and driller for Guy F. Atkinson Construction on Folsom Dam's powerhouse and a miner in the penstock tunnels.

Arriving in Sacramento, California, the closest big city to Folsom, we spent a night in a trailer park and phoned Jake Semler to discover the best route to Folsom. George Veatch had told him we were on our way. Jack had his "diggers," or work pants, but later decided to wear his regular twill chinos. Without any interstate freeways at that time, it took quite a while to reach our destination.

Jake was a good-sized, pleasant, balding man with the look of authority in his stature. His wife, Ione, had a mannish look about her and a gruff voice. She smoked cigarettes inserted into a long holder. She seemed formidable at first meeting.

They met us at the Folsom Dam trailer park, which looked quite dismal at that time with few trees and lots of sand. It was *hot*. We put porch furniture on the small cement patio in front of the trailer and sat there in the evening after the heat of the day faded. It also helped keep us from tracking the sand inside.

The park covered a few acres. More trailers were parked in the back, in areas that had some trees. A huge communal laundry room and bathrooms were at the center of the park.

One day, when I was doing my laundry and hanging it outside on the lines (they had no electric dryers at that time), several of the women were looking at my laundry and giggling. I thought they were cliquish. I asked why they were laughing. They said they wondered what my husband must look like because of the wide work pants with short legs I was hanging up. I could see their point and laughed along with them, explaining that my husband was only five-foot-four, but while working in the coal mines cut off his pant legs to avoid having them drag in the mud inside the tunnels. I met the Ramsey women that day, and we became best of friends as members of the "construction family."

The heat in the daytime was so bad we couldn't touch the outside of the trailer, but no matter how hot it was during the day, it always cooled off at night; in fact it became cold.

When Jack started work on Monday, he was surprised to see Bob McLelland, who was shift boss on that job also.

The first part of the week trailers started moving into the park as more Tunnel Stiffs arrived, looking

Above, in front of the flooded powerhouse of Folsom Dam are, from left, Pat Sweeney, and George, Sharon, and Rose Ramsey.

for work. Parking two lanes over from our trailer was Al Hoiland, nicknamed the "Big Swede," 'and his family. He was one of the most unforgettable characters I met on that job. He was a huge man, standing well over six feet tall with broad shoulders and a thatch of faded red hair. His voice was as loud as he was big. When he spoke, everyone could hear him. His wife, Emily, was a beautiful woman. She had been a widow with three children when Al married her. They had a three-year-old daughter with carrot red hair, named Kon Tiki. "Teek" was her nickname.

It took some time to become acquainted with them as Al was a private man who didn't like to visit with anyone when he returned home from work.

I don't know how he earned the "Big Swede" handle, as he was from Norway. I did enjoy visiting and listening to him. His pronunciation of V's and W's were backward in his speech.

We had been looking for a dog for Pat and found an ad in the paper that interested us. It was at a pri-

These photos from January 1953 show the flooded powerhouse at Folsom Dam.

vate home and the dogs were Boxers, registered with the American Kennel Club.

This one little pup took to us right away. She gazed at us with big brown eyes that begged, “Take me! Take me!” We bought her for $75, with her papers, and gave her to Pat as a Christmas present.

Buying that dog was one of the biggest mistakes we had ever made. Although she was a beautiful dog, she didn’t have a brain in her head. She was so stupid she would trip over her own feet. Pat was thrilled with having his own dog and loved her in spite of her ways. He named her “Beauty” and they were constant companions except when it was time to take her for walks and exercise. Somehow I got roped into that job.

Other characters we met on that job who became part of our construction family were George and Rose Ramsey and their daughter, Sharon; Virgil and Colleen Ramsey and their sons. George and Virgil were brothers. We saw them on many jobs in later years.

The Folsom State Prison is seen in the background.

Pat Sweeney fishes with Rick and Butch Ramsey on the American River near Folsom Dam in May 1954.

Chapter iv: Santa Ynez, California, 1953

Cachuma Dam

John Wiley's aerial view, shared on Wikimedia Commmons, shows Bradbury Dam, constructed in the 1950s by Guy F. Atkinson Construction Company at Lake Cachuma.

AFTER FINISHING WORK on the Folsom Dam, we moved to a park in California's Santa Barbara County, where Jack worked in 1953 as a miner on the Cachuma dam and reservoir located beyond the Santa Ynez mountain range.

Solvang, California, nestled in a valley at the foot of the mountains, was one of the cleanest, most beautiful little towns I had ever seen. It was a Danish town, with brick buildings and a huge windmill that stood out in that rugged area.

Pictured above is Cachuma Dam near San Ynez, California, in March of 1953, when Jack worked for Guy F. Atkinson Construction Company. Below is the Techolote Tunnel at Santa Ynez, California, in March 1953.

Sheepherding and cattle raising were the main industry when we were there.

We found a trailer park a few miles up the road from Solvang, in Santa Ynez, which was only a wide spot in the road. George and Virgil Ramsey had already arrived with their families. This is where we met Arthur "Whitey" and Ruth Phelps. Just five trailers made up this small park.

The Williams Trailer Park was simply a wide spot in the road.

Santa Ynez was just a rest spot beside the road, situated on a sand bank, barren of grass. Fragrant wildflowers bloomed close to the ground. A huge propane tank and a power pole behind the trailers supplied our needs.

Bob and Gerry McLelland lived farther up the road toward the job site. We were surprised to see them on this job. Bob was Jack's immediate supervisor again.

This area was sheep country. Often a car would have to stop and wait while the sheepherders drove the animals across the road, an effort that sometimes took up to an hour or more, depending on the herd size.

A school bus picked up the children, but I don't remember the school they attended.

Above is a Danish windmill in Solvang, California, seen in 2008 in this photo made available via Wikimedia Commons at http://www.flickr.com/photos/bossco/2922053648, and below is the east end of Lake Cachuma, seen in a Wikimedia Commons photo by John Wiley.

For a treat we would drive to Solvang, snoop in the shops, and look over the town. We'd sit on soda fountain chairs at little round tables outside on the sidewalk, under shade trees, enjoying ice cream treats. This is all I can remember about this area, as the job was of short duration.

Chapter v: San Francisco, California, 1953–54

Waldo Tunnel Construction

This May 1954 Guy F. Atkinson Construction Company photo shows construction of Waldo Tunnel, where Jack worked as a miner and shifter on the new four-lane approach to the Golden Gate Bridge in San Francisco.

FOR OUR NEXT JOB, we hauled the trailer to San Rafael. We were excited to see the Golden Gate Bridge, the San Francisco Bay, and the city of San Francisco.

Guy F. Atkinson Construction Company hired workers to build a four-lane highway tunnel from the Golden Gate Bridge toward San Rafael, California. Jack

Jack worked on construction of the Waldo Tunnel, seen in front of the Golden Gate bridge supports in the photo above. At left are crew members who worked on the project. Jack is seen in the center behind the man whose legs are stretched forward.

worked as a miner and a shifter on this job, driving a tunnel through the mountain under soft ground, which required installing steel supports all through the tunnel. They also drove two four-foot liner plate drifts under the double highway in that same area.

We looked for a place to live, and decided on the Golden Gate Trailer Park in San Rafael, the nearest place to the work site without the heavy traffic. The very nice park, smaller than many we had visited, was the first with grass lawns and sidewalks.

This park did not have many construction people. I think they preferred living closer to the big city. Our new neighbors were Jim and Patty Anderson from Boulder, Colorado. Jim was a serviceman stationed at the Presidio in San Francisco.

Virgil and Colleen Ramsey, who were also interested in this area, stayed with us while Virgil rustled the job. (He later decided this wasn't what he wanted). They were pleasantly surprised when they met the Andersons, who were old schoolmates from Boulder where they all grew up together. The Andersons joined our "family" and we kept in touch through the years.

We enjoyed mild weather and beautiful scenery with palm trees. A railroad track ran behind the trailer park. Many the day we heard a knock on the door from a man down on his luck, asking for food. We never turned away a hungry person. It was interesting to hear their life stories. They might have been called "tramps" and "bums," but we found them polite men traveling the rails seeking a better life. Some just liked the adventure of traveling.

Pat and I took our boxer dog for long walks across the railroad tracks for exercise. We discovered a small pond, where the dog could swim and Pat could catch polliwogs. Pat rode the school bus to a grade school in Mill Valley. I treasure these memories.

My father's brother, Jack, lived in Oakland, California, but I had never met him or his family. We

Tunnel Stiffs learned to work underground in the dark, digging tunnels, drilling holes, and building dams. These unidentified men are working on construction of the Waldo Tunnel.

contacted Uncle Jack and he was anxious to meet us and vice versa.

We became lost the first time we visited them. We stopped and I phoned Uncle Jack. “Where are you calling from?” he asked. I told him we were at a phone booth on Athol Street, although I didn’t know how to pronounce the street name. We eventually found them and shared a good laugh. Although Uncle Jack was several years younger than my father, they looked like twins. Uncle Jack’s wife, Amelia, was a very sweet lady of Portuguese descent. They made us feel like a part of their family. We visited them often and visited her

Our blue and white Oldsmobile sits at the curb as Pat Sweeney and Beauty play near our trailer in the Golden Gate Trailer Park at San Rafael in June 1954.

daughters and family across town.*(Uncle Jack was a widower when he married Amelia. They both had children from previous marriages.)

Bill (my father's nephew) and Catherine Roberts lived in the Nob Hill section of San Francisco. It was good having "blood family" nearby and seeing where they lived.

This is about all I can remember of that place in time—so long ago, with many other construction jobs to follow.

Chapter vi: Winters, California, 1954-55

Monticello Dam

Jack Sweeney, 300 feet above the Trinity River, high-scales loose rock from a mountain face while working for Guy F. Atkinson Construction Company on a diversion tunnel and shaft for Monticello Dam.

After finishing the tunnel job near San Rafael in 1954, we moved less than a hundred miles northeast to Winters, California. It was a short move in spite of the two-lane roads we traveled.

When we arrived in the small town of Winters we contacted Al Hoiland. He and his family lived in a nice tree-lined trailer park on the main street. Sadly for us, the park was full. We had to settle outside of town toward the job site ... in yet another treeless area! It was quite large, with trailers parked hither and yon, and certainly wasn't a planned park. I could sweep the hardpan ground with a broom.

We parked and settled in. Al, the "Big Swede," was Jack's immediate supervisor. They went to the job site to sign on, initially working for Guy F. Atkinson Construction Co. but in 1955, they were employed by the newly formed Gates & Fox Construction Company. This was the company's first or second job. Its crew consisted of Kirk Fox as project manager, Ed Hill as foreman, Roy Shoemaker as master mechanic, and James Clark as office manager. Other workers were George Boone, Tom McClanahan and Carrol Necaise.

For this job, Gates & Fox Construction Company oversaw construction of the Monticello Dam. Jack worked as a miner, but the job required him to high-scale 300 feet above the Trinity River while constructing the dam's diversion tunnel and shaft. They used rock bolts and wire mesh to support the tunnel and shaft.

In the trailer park, we met Miles Moore and his family, wife Pat and son Garry. Miles was another big man, who stood over six feet tall. The blue-eyed Irishman sported a head of thick, dark brown hair. He was a one-armed man, having lost his other arm in a hunting accident as a child. Strong as a bull, he could do more with one arm than most men could do with two. I saw him lift a barrel of fuel oil with his one arm to hook to his heater in the trailer.

The funnel-shaped Morning Glory Spillway, or Glory Hole, allows water to bypass the Monticello Dam on Putah Creek when it reaches capacity. It is seen at left during construction of the dam in 1957. Below is the hole in the Lake Berryessa reservoir. The dam is operated by the United States Bureau of Reclamation, which provided the photos.

I don't remember just what his job was. I know he operated heavy equipment, which was quite a feat for a one-armed man!

They were also friends of the Veatches and Semlers whom we had met on previous jobs, but were working in the Philippines at this time. They kept in touch with the Moores and us. The Moores joined our extended Tunnel Stiff family on this job.

The first week there, I enrolled Pat in the Winters grade school. A school bus service picked up him and other students.

On Halloween the school kids were going to be in a parade in Winters. Pat begged me to make a costume for him to wear in it. I dashed into town and bought orange crepe paper and a straw hat. I stitched patches on his shirt and blue jeans, and then cut the crepe paper so it looked liked straw. I sewed it into the wrists of his sleeves and jeans hems, and under the brim of the straw hat. I painted his nose, cheeks, and above his eyes red with lipstick. He won first place in the parade, hopping and skipping down the street. He wore a red nose and cheeks for about a week. I didn't notice when I painted him that the lipstick was a permanent color.

Several families who didn't follow construction lived in this place also. They had quite a few kids, so Pat had plenty of playmates.

Wayne and Zennie Hicklin lived next door with their two little daughters. Wayne worked for the electric company as one of the guys who shimmy up the light poles and put in new electrical service. They were from Blue Springs, Missouri. We enjoyed some good card games with them and the Moores. They were not Tunnel Stiffs, but joined our ever expanding family. We kept in touch, visiting back and forth when we could through the years.

Emily Hoiland took me shopping in Sacramento to buy our groceries. One day I left my car at the garage for repairs while we shopped. When I went to pick up the car, they hadn't done a thing to it. The mechanic said that when he climbed in to move the car inside, he felt something cold against the back of his neck. I had forgotten to tell him that Beauty, our boxer dog, was asleep in the back seat. He said he bailed out of there and wouldn't go back. Beauty wouldn't hurt a fly, but he didn't know that.

Another incident happened to me while I was driving back to the park one day. The roads were two lanes and someone happened to be moving a house on the back of a large truck and trailer. I pulled off the road as far as

I could, but the house was still wider than the road. It scraped the whole side of the car. That was a hard one to explain to Jack, when I told him a "house" ran into me. Pat was my witness as he was with me. (He still tells people about the time a house ran into his mother's car).

When this job ended, Gates & Fox was low bidder on a job in Glendo, Wyoming. Jack commented about how much he liked working for that company and mailed a note to Kirk Fox, asking whether they had an opening for a good miner in Glendo. Fox answered immediately and invited us to join his crew. We became permanent members of the Gates & Fox Construction Company, a family-oriented company, for the next thirteen years. They were now a growing company of eight men, well-known for completing the work well. The bigger the contracts, the more employees the company hired.

The Glory Hole at Monticello Dam is seen during low water in this U.S. Bureau of Reclamation photo.

Chapter vii: Glendo, Wyoming, 1955-56

On the Road Again

We had just left Rawlins, Wyoming,, on our way to Glendo when this photo was taken of our Oldsmobile pulling our trailer. Jack, who painted the trailer blue to match the car, was working for Gates & Fox Construction Company.

WE PREPARED FOR A LONG HAUL of nearly 2,000 miles from the Sacramento area to Wyoming. We traced the route we would take on the map, packed, and took off on a trip of several days.

Wyoming was the windiest state we had ever been in as Jack struggled to keep the car and trailer on the road. Although springtime had arrived, the weather remained cold and snow fell. We tried to spend the night in an overnight parking area but the road proved too icy to venture down the slope into the parking area.

We parked in back of a closed gas station late at night, turned on our floor heater, and fell asleep, clothes and all. Beauty slept with Pat to keep them both warm. Fumes from the floor heater made us sick but it was too cold to open any windows for ventilation.

A policeman hammered on our door in the wee hours and told us we couldn't park there. We would have to leave. What a welcome to Wyoming!

Fox *Hill* *Clarke* *Shoemaker* *Sweeney* *Fox, Jr.*

Bossing the job at Glendo Dam

A clipping shows photos of men working for Gates & Fox Construction on the tunnel and shaft at Glendo Dam, used with permission of the *Platte County Record Times.*

Kirk H. Fox of Gates & Fox Co., has been acting as project manager on the firm's $700,000 award for the tunnel and shaft at Glendo Dam. They started work March 1 and are now finishing up the job. Tunnel superintendents (walker bosses) are **Ed Hill** and **George Boone.** Tunnel foremen (shift bosses) are **Jack Sweeney** and **Tom McClanahan. Roy Shoemaker** is master mechanic, assisted by **Carroll Necaise.** The office is in charge of **James Clarke.**

October 1955

Glendo was a very small, old, western cattle ranching town with only one main street, grocery store, post office, garage, church, and school, but lots of dust and wind.

We settled in at the Pioneer Trailer Park about a mile outside of town on the plains. It was all new, with laundry and restrooms, and lots of sand and wind. I felt like we had reached the end of the world.

As we understood it, Glendo's politicians and city fathers, who had never met a construction crew, decided the rough and tough people who followed these jobs might cause trouble and upset their tranquil, little town. So they imported a house of ill repute and located it on the outskirts of town to keep the Tunnel Stiffs out of Glendo and away from their women. The "house" would also raise tax money for Glendo to install sewers and other city services.

The job site was quite a long drive from the park. A taxi service picked up the miners at their homes and transported them to work. One of the "ladies" rode in the taxi (whose operator also owned the "house of ill repute" to try to entice the men to "visit" the house). The skinny taxi driver looked exactly like some character from the Wild West, with his handlebar moustache and plaid suits.

That "house" didn't last long. The construction people objected to the town's way of thinking and took it up with the Glendo City Council. We didn't live like that. Most of the Tunnel Stiffs were married with children and did not condone having this form of "entertainment" in the community.

Of course there may be a bad apple in the bunch, but the Stiffs' way of life was good, with hardworking people simply trying to earn a living. We were like nomads, always on the move for the next adventure.

Jack worked as a shifter on this job, constructing a diversion tunnel and shaft at Glendo Dam. The tunnel was constructed in shale and sandstone. Workers erected steel supports in most of the tunnel and all of the shaft.

At the park, I met its owner, Mr. Wilson, an elderly man crippled with arthritis. He had been quite a rancher in his day and still lived in the place his ancestors had homesteaded.

I asked if he needed someone to help take care of his park–parking trailers, collecting rent, and cleaning the laundry and restrooms. I would do it for our space and utilities. That worked out well for both of us, as he had to check the park only once or twice a month.

He loved to chat and visit. I used him as a guinea pig to taste my cooking whenever he dropped in. He liked to taste new dishes and I think he timed his visits to collect his rent and electric payments so they fell on my baking days. (He didn't become a Tunnel Stiff but kept in touch with us for many years, until arthritis in his hands kept him from writing).

It was so cold and windy our first week there it blew the trailer off the blocks, with Beauty and me inside. I thought for sure it would tip over. We were lucky.

I had taken Pat to the brick schoolhouse and I was sure it was safe. We had to practically crawl across the field to reach the school because the wind was so strong. Frost lined the inside of the trailer walls as the wind pulled the windows from the walls. We sat in the laundry room to get warm.

Beauty landed in big trouble after she joined a pack of sheep-chasing dogs. The sheep ranchers threatened to shoot all of the canines when they saw them. I told them if they shot Beauty, make sure they killed her. I didn't want her left wounded someplace.

Mr. Wilson intervened with the ranchers on her behalf. Afterward, I kept her chained. I had never chained a dog in my life and hated to do it. She was okay, but when I let her off the leash to do her duty, she sped off like a bullet to join her dog pals.

Pat Sweeney poses with his beast, Beauty, in Glendo, Wyoming.

While we were in Wyoming, we visited nearby points of interest with Ed and Dot Hill. We drove to Estes Park, Colorado, for swimming and picnicking. We were thrilled to see Yellowstone National Park where the Old Faithful geyser gushing its spout of steam could be timed almost to the minute. We followed a walkway of

Above, Ed Hill cautioned Pat about the bubbling mud in the witch's cauldron; at left, Yellowstone Inn and Old Faithful; below, Pat chases a bear cub in Yellowstone Park.

wooden planks around the bubbling mud holes. One had to have very good balance to walk on them, as they were very narrow. One slip and you would be cooked.

The bears seemed tame as they sat along the roadside, waving and beckoning to visitors. We even allowed Pat to sneak up on a baby bear and pet it. It gives me chills now to think about it.

We stayed in the one and only inn–an old, wooden three-story building, that looked like a fire trap. Our rooms were on the third floor, while the first floor was the dining and observation area.

We met the Ed Hills, Roy and Bertha Shoemaker, and Bob and Eloise Richardson on this project and added them to our family of Tunnel Stiffs.

One day we womenfolk decided to shop in Cheyenne and see what the big city was like. The city looked old, like the western cities in the movies, with a huge mercantile store featuring a huge safe embedded in an aged wooden floor, with a dial to open it flush with the floor.

Pat was really interested in the safe and I let him sit and play with it while we did our shopping. I thought that would keep him busy and not underfoot.

It wasn't long until a store clerk came looking for us, hauling my redheaded, freckle-faced kid, bellowing, "Who does this kid belong to?" I thought, Uh-oh. What did Pat do now? He had played with that safe dial and opened the safe. The clerk said that no one person knew the combination and it took three of them to open it, as each person knew part of the combination. I asked Pat about it and he said he could just feel the tumblers moving while he was playing with it... The owner sure bawled out my boy.

When Pat started opening mailboxes with dials in the Glendo Post Office, he got kicked out and told not to come back. At that point, I thought, My gawd, my kid's gonna be a safecracker!

From top, the Pioneer Trailer Park in Glendo, Wyoming, after a blizzard in March 1955; Pat and Beauty on the way to school with Glendo, Wyoming, in the distance; and the laundry and restrooms after the March 1955 blizzard.

Mrs. Downey, who was the wife of a town politician, ran the post office too. She was the mother of Eloise Richardson who became one of my dearest friends. Now when we reminisce with our friends, they always remember that event.

This brochure from the 1950s, used with permission, shows the development of the Glendo Dam in Wyoming.

HISTORIC GLENDO

BY VIRGINIA COLE TRENHOLM

When Robert Stuart and his Astorians plodded eastward on the icy surface of the North Platte River in 1812, they little realized that they were charting the course for the great westward migrations which were to take place in their lifetime. They were chiefly concerned with escaping from the Arapaho, who had frightened them from their winter quarters on Poison Spider Creek (near Casper).

Trappers and fur traders had established the natural roadway along the Platte before Horace Greeley made his celebrated statement: "Go West, Young Man". With this slogan ringing in their ears, 1,000 recruits joined Marcus Whitman's wagon train, which ultimately saved Oregon for the United States. It was without doubt the largest caravan ever to wend its way along the Platte. It followed in the wake of John Charles Fremont, "the Pathfinder", who compiled data for the first travelaid of the West, the year before.

Of the thousands who streamed through this area, only one stayed to make a study of the native Indians. Francis Parkman's account of the Sioux has become a classic—**The Oregon Trail.** Called the "Great Medicine Road of the Whites" by the Indians, the road to Oregon assumed the name Mormon Trail in 1847, during the Mormon exodus.

The farsighted Saints tarried in the Glendo area long enough to improve roads and build the first of their five mail stations joining Fort Laramie and Salt Lake at Horseshoe Creek. The definite location has never been found, but blueprints show that the settlement amounted to 160 acres (with irrigation ditches, probably the first in Wyoming) inclosed by pickets. The buildings were burned by the Mormons as they fled toward Salt Lake before Albert Sidney Johnston's advancing Army for Utah.

During the Gold Rush days, the Emigrant Road along the Platte was called the California Trail. Indelible ruts show an ever-widening trail as excitement reached a fevered pitch. Each tried to out-distance the other in his mad race to "the Diggings"

It was not until travel had slowed down to a more normal pace that the Overland Stage Line was established by Russell, Majors and Waddell, and the roadway became known as the Overland Trail. Stations sprang up along the route, with one of the most important at Horseshoe Creek. It was manned by the notorious Slade, who is said to have carried his adversary's ears in his pocket after an altercation with Jules Reni.

A local legend has it that a lone New Yorker held up a stage in the Horseshoe vicinity and was forced to cache his loot ($30,000 in gold) among the rocks on the side of Sibley Peak, which overshadowed the station. Many a "prospector" has dug on the rugged slopes.

One of the most celebrated guests at the Overland Stage, Pony Express and Telegraph Station was Mark Twain, who gives an account of "Alf" Slade in **Roughing It.** Slade is presumed to have had a hideout on the east side of the Platte, not far from the Glendo Dam, where the "Slade Chimneys" indicate that a building may once have stood. It was here that he and his gang were presumed to have preyed upon the emigrants following the opposite branch of the trail.

Slade's wife, Virginia, saved the Stage Station from being burned while she was living there. The Overland employees got out of hand, after burning out E. W. Whitcomb's establishment nearby. They dumped one barrel of his liquor in the well at the station in order to have a perpetual supply and proceeded to drink the other. While in a high mood, someone suggested that they burn the Stage Station. Virginia drew a revolver and threatened to shoot the first to make a move. By the time Slade arrived, they were sober. She remained loyal to him until his death, at the hands of the vigilantes in Montana.

Buffalo Bill, while stopping at the Overland Station, made a hunting excursion up Horseshoe Creek. There he encountered a gang of horse thieves, from whom he barely escaped with his life.

After the Overland traffic was re-routed to the Cherokee Trail in Colorado, 1862, the telegraph station continued in operation. It was here that Portuguese Phillips stopped to send a message to Fort Laramie during his famous ride down the Bozeman Trail from Fort Phil Kearny in 1866. Not trusting his message alone, he rode on to "Laramie", to gain the distinction of being Wyoming's Paul Revere.

In 1868, Crazy Horse and about 60 Sioux warriors attacked old Horseshoe station and burned it to the ground in a three-day battle which stretched from Sibley Peak to the hills south of the Cassa Flats. The peak, where the Indians were first encountered, was called Sibley because it resembles a Sibley tent used during the Civil War. It is one of the well known landmarks along the Emigrant Trail.

Near this peak, the McDermotts operated a stage station and post office, called Bellewood, between 1887 and 1892. When the railroad was built connecting Wendover and Orin Junction, the post office was moved to the flats and was given the name Glendo, presumably from the glen at Bellewood where the crew was encamped. Upon construction of the railroad, the main street was graded, lots were sold and the new town was on the map.

SIGHTSEEING IN ABUNDANCE IN THE GLENDO AREA . . . .

In addition to the many specific points of interest in the Glendo area, there are a number of nearby places to visit for camera "bugs" and sightseers.

Our Medicine Bow National Forest offers the visitor a wonderworld of nature on display. Easiest access to this area is gained by traveling the "Ridge Road" which runs due west from Glendo, crossing beautiful rolling plains before reaching the mountains. At Esterbrook, 22 miles west of Glendo, one finds the Church of Our Savior set in a location so undisturbed and beautiful that the sight of it alone makes the trip a worthwhile venture.

One mile south of Glendo the Horseshoe Road leaves Highway 87 and winds its way westward through one of the most scenic mountain valleys in the West.

Wyoming's VACATION PARADISE

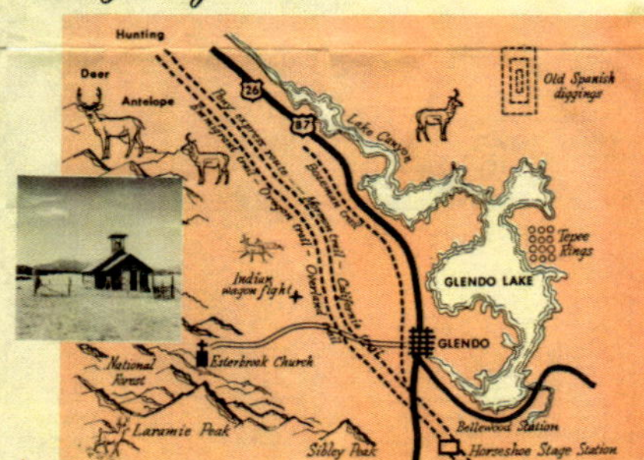

POINTS OF INTEREST

Spanish Diggings — Oregon Trail — Mormon Trail — Emigrant Road — Bridgers Ferry — California Trail — Overland Trail — Horseshoe Stage, Pony Express and Telegraph Station — Bozeman Trail — Bellewood Station — Pony Express Route — Wagon Fight Site — Sibley Peak — Esterbrook Church — Glendo Dam and Reservoir — Laramie Peak — Lake Canyon

Chapter viii: Oakdale, California, 1956

Tulloch Dam

This scene shows the blasting for the final "holing through" process on the Tulloch 550 foot diversion tunnel.

This photo was taken inside the Tulloch diversion tunnel as workers prepared to dynamite for the final "holing through."

These newspaper clippings show the dynamiting of the Tulloch diversion tunnel in 1956 by Gates & Fox Construction Company.

AFTER WE FINISHED WITH THE GLENDO DAM project, we packed again and moved west to Oakdale, California, where Jack worked in 1956 as a shifter building a diversion tunnel on the Tulloch Dam.

This was a short job and I don't remember much about it. The men didn't use any extra ground support in construction of this horseshoe-shaped sixteen- by sixteen-foot tunnel.

We became acquainted with the Richardsons and Shoemakers. The Richardsons lived in an apartment in Oakdale.

Roy and Bertha Shoemaker lived about a mile from us, on the other side of the road in a nice trailer park. I visited them often. They were from Fort Collins, Colorado. Roy, the master mechanic on this job, was a tall, slim, pleasant man who always had a story to tell. He loved to share funny jokes.

His wife, Bertha, a homemaker, was on the heavy side and a very sweet lady.

She did all of her cooking from scratch. They were in their fifties at that time and I thought they were "old." In all the time we knew and worked with them, I never once heard either of them gossip or run down another person.

I often met Eloise "Tilley" Richardson and her two boys in town when I did my shopping. They were originally from Glendo, Wyoming, and decided to join the Tunnel Stiff group, working for the Gates & Fox Construction Company.

I don't even remember the name of the park where we stayed, but it was the only one that took pets and we had our humongous, lanky "idiot," Beauty. I believe we were the only Tunnel Stiffs in the park. We met, liked, played cards, and visited with several of the inhabitants… Some of them lived there permanently.

Our neighbors, Bud and Lee Riggles, fell in love with Beauty. They took her for walks and fawned over her. She spent so much time with them that we gave her to them. They took her to their ranch, where we hope she lived a good, unfettered life. The Riggles fell in love with Pat too, but we wouldn't part with him. Pat had started school and didn't have the time to spend with Beauty. We discussed it with him and he agreed she could live with them. He could play with the dog whenever he wanted. They sold her just before we moved from there.

While working on the Tulloch Dam project, we sold our trailer and bought a "mobile home," which is what they were called by then. The dark pink and white Roadliner, which was thirty-two-feet long and eight-feet wide, could be expanded electronically to a width of thirteen feet. This very comfortable home seemed more like a normal house on the inside. I especially liked the kitchen across from the dining room, with its eight-foot-wide galley and built-in copper appliances—stove, wall oven, and refrigerator. I owned a set of dishes with a chicken motif and, copying the plates, I painted chickens to match on the kitchen cupboards. That brightened it up. The back door led directly off the kitchen, and another back door opened off the bedroom area.

Our new mobile home contained one of the first slide-outs, which was expanded electronically to give us more room. Our trailer is seen in Dodge Park, Oregon.

Pat's narrow bedroom featured a trundle bed, which could be pulled out from under his twin-size bed. A closet, chest of drawers, and a small stand stood by the bed.

The day we opened and set it up for the first time, we entertained our first visitors! Jake and Ione Semler and Miles and Pat Moore decided to drop in unannounced for the weekend. (Our friends were always welcome, even on a surprise visit). We had plenty of room for them. The Moores slept in Pat's room and we gave the Semlers our bedroom. Jack and I slept on the couch in the living room, with Pat slumbering on a pallet on the floor.

After the "holing through" on this job, we left for Lewiston, California, as Gates & Fox had a new job tunneling for the New Trinity dam. The company moved the trailer for us because it was too heavy to be towed behind the car. The company would move the oversized or overweight trailers of its permanent employees.

We traveled with the Shoemakers and slept and ate with them in their trailer.

Chapter ix: Lewiston, California, 1956–58

Trinity Dam

Jack was a walker boss for Gates and Fox Construction Company as it built the diversion tunnel and shaft at Trinity Dam in Lewiston, California, drilling through blocky granite to install steel supports.

OUR FIRST NIGHT ON THE LONG ROAD from Oakdale to Lewiston, California, we encountered one of the worst rain storms ever. Rain poured down in sheets. Roy Shoemaker couldn't see two feet in front of his truck.

Driving at a snail's pace, he spied a restaurant. We decided to park there for the night and eat our supper and breakfast before we started again in the morning.

Trinity Dam as seen in 1991, nearly forty years after Jack and the Tunnel Stiffs worked on the project.

We slept in their trailer that night and inhaled a quick breakfast in the restaurant. What a good change from having to cook while in transit!

It was still raining, but not as bad as the previous day. We hit the road again for Lewiston.

When we reached the city of Redding, we spent the night there. Lewiston was fifty miles north, uphill. I think it was the longest fifty miles we drove.

Once we left the main highway, we bumped along a muddy two-lane road. We slid and slithered all the way up the mountain, and stayed with the Shoemakers until the company hauled our trailer to the Trinity Trailer Park in Lewiston. Workers drove our trailer into the park, slipping and sliding all the way, and parked it where it stuck in the thick oozy mud.

The men had to place planks on the mud to maneuver around the trailer, opening it to restore order inside, making it livable again.

CAMP SITE type of area in the Trinity River valley, now occupied by trailers housing employees on some of the preliminary contracts. Pipe on suspension span is sewer outfall.

DIVERSION TUNNEL around the Trinity Dam site was driven under separate contract. Truck mounted jumbo with three decks was used to drive 2,280-ft. tunnel.

These newspaper and newsletter clippings describe the Trinity Dam project and the role of Kirk H. Fox, a partner with Clay Gates in Gates & Fox Construction. Used with permission from the Trinity *Journal*, Weaverville, California.

KIRK H. FOX is partner and field manager of Gates & Fox, the outfit that drove the diversion tunnel for Trinity Dam. This $1,348,000 job is an important preliminary to the Trinity River project of the Bureau of Reclamation which is described starting page 72 of this issue. Gates & Fox of Burlingame, Calif., holed through the diversion tunnel late in February. Their operations will be described in an article to appear next issue.

The Gates & Fox office trailer and the Shoemakers' trailer in Lewiston. Below, our new mobile home and car seen in the Lewiston Trailer Park, which was built around us after we settled in the muddy park.

We finally bought a "swamp cooler," which was installed on the trailer roof to provide air conditioning. We operated it with a thermostat inside. What a joy to be out of the heat! We could keep the windows closed, which cut down on dust inside. We also bought a television set, small by today's standards, with "rabbit ears." The antenna worked and we kept up on the entertainment and news, squinting through the "snowy" screen.

We never locked our doors. Many the time we would return home to find our trailer with a single man or two as visitors," watching television, drinking beer, or just resting in our living room. We owned the only place large enough for the guys to gather, unless they stopped at the hotel. I think the hotel grew old to them after a while and they had to buy drinks when gathering there.

We even had a house guest in Lewiston. The Hills had lost their trailer in an accident; the wind was so strong it twisted the trailer from the hitch behind truck and slid over a cliff and was demolished on the way to this job. Thankfully, Ed Hill didn't get a scratch and was able to drive the truck to our place.

Ed's wife, Dot, and daughter Joyce stayed in a motel in Grants Pass, Oregon, until their new trailer arrived in Lewiston. Ed lived with us until his home and family arrived. Living in Lewiston seemed to me like the end of the world with a lot of pine trees protruding from the mud. The park had no other trailers in it at this time. When the mud finally dried, the whole park grew sandy and dusty. Rob Roy Neilson, owner of the park, drove a huge water truck, and sprayed the roads to keep down the dust.

Lewiston was a small and ancient town (if it could be called a town). It boasted a very old two-story hotel, with room and board for visitors. It even smelled old and musty. The round bar sat in the center of the room, surrounded by chairs and tables. The main floor sported a good-sized poolroom. Mabel Coty worked as operator in Lewiston's old-time phone service. We had to ask for "Operator" to get Mabel on the line before calling someone. Mabel's daughter, Rae Gray, and her husband, Mason, operated the hotel.

The town featured one gas station and a very old grocery store stocked with the bread, milk, beer, sandwich meat, cereal, and coffee—the necessities folks needed without having to drive the fifty miles to Weaverville or Redding for their staples.

The one-room schoolhouse looked like it had been there forever, with the scratched desks where countless

LEWISTON SWITCHBOARD — Mabel Coty and her ancient telephone switchboard have been giving Lewiston residents 24-hour phone service for more than a year. Both soon will be replaced by a dial system to serve the growing Trinity River project area.

Mabel Coty, the telephone operator for Lewiston,California, was featured in the local newspaper in 1957, the *Trinity Journal*, which gave permission to reprint this article.

students had carved their names. It had been built in 1865, but when the Bureau of Reclamation started work on Trinity Dam in 1957, the student population grew so much that a new elementary school was opened in January 1958.

From town we crossed a long bridge over the Trinity River to reach the trailer park. On the park side of the river sat an old post office and a beer joint. I didn't see any houses in the vicinity except one on the hill that belonged to Rob Roy Nielson and his family, who built the Trinity Trailer Park. When we finally met some of the local inhabitants, we discovered several still had dirt floors in their homes; a few had never ventured more than ten miles from home. Unbelievable!

Several other contractors worked on this job, with their employees living in two other trailer courts closer to the job site. I met Kirk Fox on this job. Another tall six-foot-plus man, Kirk didn't walk; he lumbered, with a slow, sure gait like a big old bear.

I found the balding, thirtyish man to be a very nice person. His wife, Dawne, and four children lived at their home in Grants Pass, Oregon.

The Richardsons, who also worked on this job, lived in a house in Douglas, a small town, quite a drive from Lewiston, until they could find a place in Lewiston.

Gates & Fox hired John Lindquist as engineer. His family, wife "Boots" and daughter Mary, lived in an ancient two-story house past the Lewiston Hotel.

Eventually they bought a two-story mobile home—the first and last two-story trailer I've ever seen. It featured a large kitchen, bathroom, and living room with a bedroom upstairs. They parked alongside the rest of us.

We also met Norwood Driscol and his family, wife Marie, and two sons on this job. Norwood, deaf from a childhood disease, was a first-rate master mechanic who could feel the vibrations from the equipment and knew what needed to be fixed. He also could lip-read and understand sign language. He spoke, but in a low guttural voice that was hard to understand until we knew him better.

Not only was Norwood a good dancer, he could keep dancing long after the music stopped. He was no wallflower in the dancing department!

Marie Driscol, a somewhat heavy woman with snapping, big brown eyes and a ready laugh, was a joy to know. We became like sisters over the years.

Roy Knutson, the office manager, originally lived in Ames, Iowa. A very handsome young bachelor, he lived in Redding, California, and drove to work and back every day. When we knew him better, he often stayed with us. He swapped cars with me when he had a date,

Above, I am seen in front of the Lindquists' double decker mobile home with my little Nash Metropolitan car, which was cute but proved challenging with its unfamiliar stick shift. The Richardsons moved into the house after the Lindquists left. At right is master mechanic Norwood Driscol in Lewiston, California. He was the first person to drive across Death Valley on a motorcycle.

as we owned a new Nash Metropolitan. I guess it impressed the girls. He later met and married Mary Sainsbury, Ed Hill's sister-in-law from Canada.

The Bob Sheppard family moved in behind us. Bob also worked as a mechanic. Pat greeted them, saying, "Hey, you should meet my mom. She has red hair, false teeth, and is a lousy cook."

Bob's wife, Frieda, told me about Pat's greeting when I finally met her, Bob, and their children. Frieda wasn't in the mood to meet anyone at that time, having just lost their youngest son to a heart defect. It took awhile but we finally became life-long friends.

Miles Moore worked on this job, too, and he moved with his family into this trailer park. It was good to see these old friends again. We drove over a narrow, winding road across the mountain to Weaverville, several miles from Lewiston, to do our shopping. Weaverville also provided trailer parks in the Trinity area with laundry and dry-cleaning services. What a boon that was. They fixed every little tear in a garment and replaced missing buttons. We also drove that long winding road to Redding to buy clothing and appliances. We would have a good time shopping and eating lunch and dinners in the "big city."

I found a job of cleaning the restrooms in this park also. That paid for our parking space and utilities and put a few extra beans in our pot.

The men's restroom was at one end of the park and the women's on the other. I started at 5 AM, after the men left for work, and finished cleaning and disinfecting both before noon.

The Neilsons asked me to clean their house too, which I did for a while. It was a huge three-story house and a bit too much for me.

The editor of the Weaverville *Weekly Trinity Journal* contacted me to see if I would be interested in writing a weekly column for his newspaper, covering all three trailer parks, and the several contractors who worked

News by a Dam Site

by Sally Sweeney

EDITOR'S NOTE

To serve the needs of the expanding Lewiston area, in addition to our regular correspondent, Viola Mead, the Journal now has a new reporter in the area whose duties will be primarily connected with the project and its workers.

A young housewife, Sally Sweeney is the name of the reporter, and with her husband, John, and son, Pat, makes her home in a trailer at the Lewiston Trailer Court.

Following construction for the past eight years, her husband is a "walker boss" on the diversion tunnel job for Gates & Fox.

Mrs. Sweeney hails from Rome, New York, while her husband, was born in Kingston, Pa.

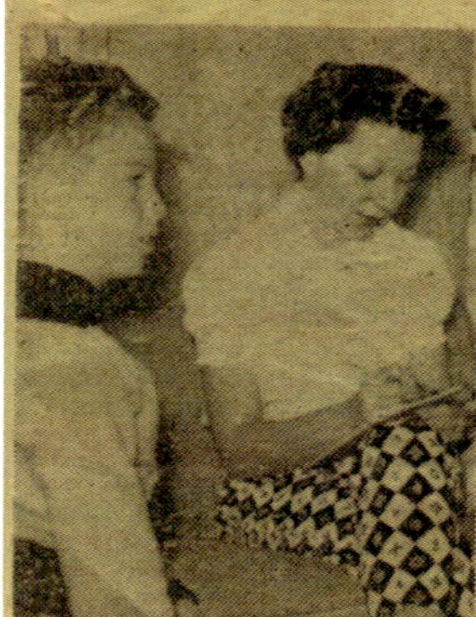

SALLY SWEENEY, and her son Pat, 11 year old 6th grader. Mrs. Sweeney is Trinity Project reporter in the dam site area. Pat will deliver papers in the Trailer Court.

Visitors to the community include: Mrs. Marguerite V. Neilson of Oakland, who is visiting her son and daughter-in-law, Mr. and Mrs. Rob Roy Neilson.

Mrs. Lyle Garrison of Weaverville is visiting her husband for a few days. She plans to move here soon. They were entertained by Mr. and Mrs. Harvey Arbuckle. Garrison and Arbuckle run the local batch plant which supplies concrete to the government housing project.

Mrs. Mamie Gray of Redwood City is visiting her son and daughter-in-law, Mr. and Mrs. Mason Gray.

Richard Johnson of Eureka is visiting Mr. and Mrs. Lyle Garrison.

Some of the lucky ones who got away for the holidays include: Mrs. Eva Copeland, fifth and sixth grade teacher, who is visiting in Mexico.

The Cecil Santos family are also visiting in Mexico.

Mrs. Freida Sheppard and children are spending a few days in Napa, Calif.

Jean and Roy Conner, managers of the local grocery store, spent their Christmas vacation in Los Angeles.

Mr. and Mrs. Miles Moore and son Gary spent the weekend in Fairfield.

Rev. and Mrs. Lindon D. Real are vacationing in Florida for five weeks. Rev. Real is the pastor of the Community Congregational Church in Lewiston.

Mr. and Mrs. Ed Hill and family left Saturday to visit relatives in San Diego. They are entertaining Miss Mary Sainsbury, from Canada, during the holidays.

on this dam. I did that too. The editor, who was quite the "dude," looked like someone from the Alps, wearing an Alpine hat with a little feather and a billy-goat goatee. He knew he would sell a lot of papers with a local reporter. Everyone liked to see their names and read the goings-on in the other parks. I was paid per inch of space in my column, called News by a Dam Site. Although I was one busy gal, I loved interacting in public and meeting new people.

Pat's career as a paper-boy was short-lived. He didn't mind delivering the paper but wouldn't ask for his payment when it was due.

As our Gates & Fox crew grew better acquainted with folks in the park, we decided to hold monthly meetings and luncheons and invite them. We also invited local people who had lived around there all of their lives when we met them. They were interesting;

Above, Bob Richardson and Jimmy Hastings shoot the rapids while hauling supplies for the new footbridge over Trinity River near Lewiston, California, in May 1956. At left and below, Jack works on the pile driver as employees build a temporary span to replace the bridge washed away by flooding.

we learned a lot from them.

Cattlemen from the area held their cattle drive in the fall, herding their cattle, which had been roaming free in the mountains, to their campgrounds for branding and castration of the bulls.

They built huge campfires to keep their branding irons hot. They called the bull testicles "mountain oysters," which they roasted over the campfires and nibbled as a ranchers' delicacy. We were invited to partake in sampling their food, but we didn't have the stomach for it. The thought of smelling burning hair and flesh and roasting the testicles over the hot coals seemed so barbaric.

When a big flood washed away the bridge over the river, anyone parked at the work site had to wait for

TRINITY FOOTBRIDGE—A small footbridge spans the Trinity river at the opening of the Gates and Fox diversion tunnel, replacing a larger bridge which was swept downstream during high crest of the rain-swollen river earlier this year. Two other bridges have recently been put up on the river by the Trinity Dam Contractors.

TRP brings new bridges

The Trinity river is now bridged at four places between Lewiston and Minersville, thanks to the Trinity River project.

One of the bridges is the old Lewiston bridge, repaired after the Christmas floods of 1955. The other three have been put across the river by Trinity project contractors.

The two newest bridges have just been completed by the Trinity Dam Contractors. One crosses the river at the construction yards near the dam site, the other crosses about three miles upstream from Lewiston, and connects with the existing road running from Lewiston to the dam site.

The Lewiston bridge will not hold the huge loads of equipment which will pour into the dam site, so the TDC will haul the equipment up the east side of the Trinity river along the Clear Creek tunnel access road, then cross on the Bailey-type bridge three miles from Lewiston.

The fourth bridge, now suitable for foot traffic only, spans the river upsteam from the dam site at the inlet portal of the Gates and Fox diversion tunnel. A former bridge there was washed out this winter during high water, and has been replaced by a foot bridge.

Flooding in May 1956 destroyed bridges over the Trinity River, including the old bridge between the main part of Lewiston and our trailer park. Used with permission from the *Trinity Journal*.

Pat Moore and Jake Semler visited the old Lewiston Hotel, in the early 1990s.

construction of a new bridge to retrieve their cars. It took awhile for Jack to drive our car home again.

Water didn't flood our trailer park or Lewiston, but came mighty close, lapping at the bridge between our park and the Lewiston Hotel and grocery store.

Gates & Fox held a "holing-through" party in Douglas City with dinner, drinks, and dancing. We dressed in our best "bib and tucker"; the men put on ties and polished their shoes while the ladies wore long gowns.

In 1958, with Gates & Fox between jobs, Jack worked as a shifter for the Shea Construction Company in the Clear Creek Tunnel near Whiskeytown, California. The job entailed building an eleven-mile tunnel. The outlet end sat on very poor ground so they used steel supports for the 4,000 feet where he worked.

We had already moved on to our next adventure in Winnemucca, Nevada. I believe it was in 1959 or '60.

Kirk Fox and Ed Hill, above, and crew members, seen below, were elated after finally holing through on the Trinity Dam project. At left is the dedication plaque for the Trinity Dam. President John F. Kennedy visited the site and dedicated the project.

Chapter x: Winnemucca, Nevada, 1957-58

Mining for Cinnabar

The Cinnabar mine and bunkhouse in Winnemucca, Nevada.

AFTER THE TRINITY DAM JOB ENDED, Gates & Fox sent three men—John Lindquist, the company engineer; Bob Richardson, another walker, and Jack—to a mine near Winnemucca, Nevada, to prospect for cinnabar, a red mercuric sulfide.

They set up a percussion drill liner in the mine at different levels and drilled 125-foot holes, panning the cuttings as they drilled. When they found color in the drill cuttings, they would drive a small drift or rise to that area. They also used a retort—a chamber for distilling or decomposing substances using heat—to process the ore. This project lasted for four months.

Since our trailer and the one owned by the Lindquists were so heavy, we couldn't possibly haul them up the rocky, rutted road to the mine. We saw an area with green poplar trees that looked like an oasis in the desert, halfway between Winnemucca and the mine, quite a ways off the road. We met the people who owned the land and asked if they would permit us to park on their ranch.

It proved to be the Case family homestead owned by Bill and Reva Rice. Reva inherited the ranch from her family. She and her husband, who had lived there several years working as ranchers, were happy to rent us an area for our trailers. They lived a very lonely life in the middle of nowhere and welcomed our company. The men and Pat helped with their haying and other ranch chores.

At left, Jack Sweeney is in foreground as Pat hangs on the back of the hay wagon. Below, they are stacking the hay after harvesting it.

We parked by the bunkhouses not far from the main house. To our surprise, the bunkhouses had a flush toilet, with a huge tank above it providing the water for flushing. We just had to watch out for rattlesnakes, scorpions, and huge black spiders when we used the toilet as it was a favorite place for snakes seeking a shady spot.

A noisy, gas-powered generator provided the electricity. Because it was expensive to run, it could be started only at certain times—to run the fridge a few times each day to keep it cold, and for cooking and showering. The poplar trees provided shade and we enjoyed our stay there.

Pat Sweeney is at the far left, seated on the horse with Allen and Edna Rice in front of the old Case homestead.

The Rices' daughter, Edna, about thirteen years old, was afflicted with a twisted spine and humpback, making it hard for her to walk. She wouldn't allow anyone to comb her hair or do things for her. Her family adored her and carried her to the fields when they had chores to do.

Her brother Allen was about four years old and Edna would tag along after him. The two of them would furtively watch us through the gate; it was a novelty to see other people and kids. Eventually they mustered the nerve, with a little coaxing from us, to visit us. They loved it when we held evening picnics with a bonfire to roast hot dogs and marshmallows. At times the Richardsons would drive down from the mine to picnic with us.

Driving in the desert, alive with jackrabbits, proved hazardous for the animals, especially after dark. They froze in the headlight beams of cars and proved impossible to avoid hitting. The dead, bloated bodies of the rabbits littered the roadsides.

The Richardsons moved into the living quarters of the bunkhouse. Anyone else who stayed at the mine site during the week and visitors also stayed in the bunkhouse. Eloise, who was known as "Tilley," cooked for the crew and Gates & Fox visitors. She lived a very lonely life as the only woman at the mine. She said she felt like she was living on Pitcairn Island (from *Mutiny on the Bounty*).

Often the Richardsons met us in Winnemucca and we stayed in a motel for the weekend. We could stop at a park, swimming pool, movie theater, or several cafes with slot machines, and shop at a good-sized grocery store.

Usually we swam and picnicked in the park. Someone took the kids to the movies. Around suppertime we ate and played the slots in the cafes. Those weekends proved a nice restful time and break from mining.

I worked a job at this site too. Kirk Fox asked if I would take ore samples to the assayer, who lived over the mountains and somewhere in the desert valley, I'm not quite sure where. Climbing into the small Nash Metropolitan, Pat and I took a long, winding drive into the mountains, which was like following a trail, to reach the assayer's office. I was lucky to find the place.

Pat and I looked forward to those trips, traveling "off the ranch" every two weeks. When Pat was twelve years old, I let him drive the route once. I thought, What could it hurt, being way out in the middle of the desert? Our small car didn't hold the road very well, leaving Pat scared out of his wits. He slammed his foot to the floor on the gas pedal and froze. We zigged and zagged back and forth across the road and through the desert. Thank goodness it was a flat area with no one else on that dirt road. I finally grabbed the keys, shutting off the engine, and we came to rest against a sand bank. I didn't think he would ever drive again. I know I wouldn't give him a chance to try with me in the car.

I kept in touch with the Rices for many years after we moved. Bill and Reva sold the ranch and moved to Elko, Nevada. Eventually, Edna entered a children's home, which broke Bill and Reva's hearts, but to their surprise, Edna was very happy and content there.

The last time I heard from them Edna had died. Bill and Allen had attended an air traffic control school. Both worked as controllers and loved their jobs.

John Lindquist, left, and Jack Sweeney are seen leaning on equipment in Winnemucca, Nevada.

Jack Sweeney, seen at right, and Bob Richardson collected mercury from the melted ore from the retort.

When this job ended, we headed back to Redding, California. The Richardsons bought a new trailer. We parked in the Cold Creek Mobile Home Park on the outskirts of Redding, California.

Chapter xi: Sandy, Oregon, 1958

Bull Run Dam

Ed Hill and Jack Sweeney grouting off a tunnel face from the surface to resolve an "impossible situation" at the Bull Run Dam near Sandy, Oregon. After firming the opening, workers drilled the tunnel through the grouted area.

OUR NEXT JOB TOOK US to Sandy, Oregon, where Jack worked as a walker boss on construction of two diversion tunnels at Bull Run Dam. Both tunnels were driven with liner plate for support. One tunnel was eighteen feet in diameter; the other was nine feet. This dam, constructed to provide domestic water for Portland, Oregon, required considerable grouting in the 18-foot-diameter tunnel to hold the ground. Workers did most of the grouting on the center line of the tunnel from the surface, drilling holes with percussion drills.

Gates & Fox built their own trailer park for this job, calling it Dodge Park, a short distance from Sandy.

The men leveled a hillside near the main road to the job area. Each family had huge parking spaces with graveled lanes connecting the spaces.

Besides us, the Hills, Shoemakers, Richardsons, and Driscols parked here. A rather large family lived in a small house at the park entrance. A very small store sat across from a state park.

Tall grass and weeds stretched from our parking space almost to the park entrance. I pulled the weeds, chopped the grass, and created a nice lawn and space for a flowerbed.

Jack had wanted a hydroplane boat for a long time, as we often lived near boating areas. So he took over my big yard, erected a frame, and covered it with canvas, then sent away for directions to make a small Punkin' Seed boat. (There went my nice yard!) Our place started to look like "trailer trash" for sure.

The boat, almost finished, sits near our trailer.

What a thrill ... it works! Jack steers the boat on a lake near Mount Hood in Oregon. Below is our new Thunderbird carrying Jack's little Punkin' Seed hydro, which I painted.

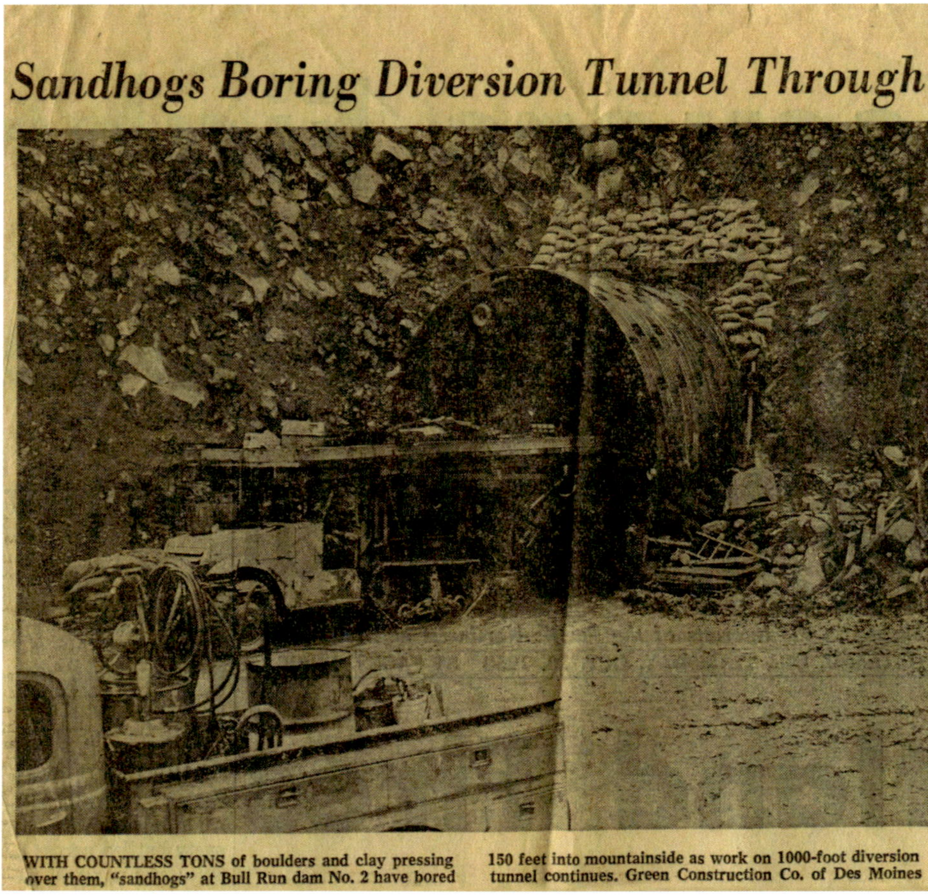

Sandhogs Boring Diversion Tunnel Through

WITH COUNTLESS TONS of boulders and clay pressing over them, "sandhogs" at Bull Run dam No. 2 have bored 150 feet into mountainside as work on 1000-foot diversion tunnel continues. Green Construction Co. of Des Moines

This *Oregon Journal* clipping, used with permission, shows the Bull Run dam project.

Jack and Pat worked on the boat on weekends and during any free time. They built the little beauty! I painted huge dragons down each side and wrote AZADRAGGIN in Chinese lettering on the back side.

Roy and Bertha Shoemaker helped Jack christen his Punkin' Seed hydro in Hoods Canal, in Oregon.

The state park employed a guard who sat in a little building and drank, often falling asleep. Pat, Jim Bob Richardson, and some local boys enjoyed swimming in the river at the park. They waited until the guard fell asleep, then copped his cigars and smoked them. They swiped candy bars at the grocery store, and after we

Mountainside for Bull Run Dam No. 2

—Photos by Bill Bailey.

is doing job on $3,345,000 contract from city of Portland. Deadline for finishing 19-foot wide steel-sectioned tunnel is next June 1. Fresh air has to be pumped through pipe at top of tunnel which is 25 miles southeast of Portland.

A miner was killed during an accident on this job.

overheard the kids bragging about this escapade, we immediately contacted the store owners who already knew what the kids were doing. They told us they placed old, wormy candy where it would be handy for the boys. The store couldn't sell them anyhow. They pulled a joke on the kids.

The town of Sandy was small and very nice. Quite a few Tunnel Stiffs rented apartments and motels while they worked on this job. We met Lee Kennard and his wife, "Pete," and their family, as well as Wayne and Merle Cornett and their family. They worked on several construction jobs with Jack.

Pat Sweeney is seen above with Oregon's Mount Hood in the background. He is the tall, lanky boy in the striped shorts in the photo at left, where he is seen with other kids from the Dodge Trailer Park. From left, are Gary, Pat, Mary, an unidentified boy, and Jim-Bob Richardson.

I remember the Fireside lounge, a few miles outside of Sandy, served mighty fine food and drinks and provided a dancing area. We often drove to nearby Mount Hood on special occasions. It was beautiful! We enjoyed eating wonderful food at the lodges there.

Chapter xii: Feather River, California, 1959-60

Oroville Dam

Jack worked on the relocation of a railroad and construction of two tunnels as part of the Oroville Dam project on the Feather River in California.

JACK WAS A WALKER BOSS on the next job, working for Gates & Fox Construction Company on construction of two tunnels and relocation of a railroad for the Oroville Dam project on the Feather River in California.

The job required a top heading twenty-one feet high driven in both tunnels. Most of the tunnels were supported with steel supports placed on wall plates and rock-bolted to the ribs. Eleven feet of bottom was taken up after holing-through. Posts were placed under the wall plates in bad ground areas where rock-bolts would not tighten up.

Building two tunnels at the Oroville Dam in California during the late 1950s proved to be the most dangerous job we worked on, as evidenced by the first-aid station and ambulance parked at the foot of a hill below the park, where almost daily someone suffered injuries serious enough to warrant a trip to the hospital.

When we arrived in Oroville, the Hoilands and others we knew were there. We decided to move to a park closer to the job, which required a long, scary drive up the Feather River Canyon about fifty miles from Oroville.

The drive took us straight up the mountain's cliff side on a two-lane road—another dangerous road without guard rails. The higher up we drove, the smaller the Feather River looked, resembling a silver ribbon snaking its way through the valley below us.

I don't remember how long the drive took, but we finally reached the Feather River Trailer Park. The scent from huge trees of cedar and pine permeated the cold air. This park, dark because of the trees, was like living in the woods.

We had no sooner parked and blocked up the trailer when the Tunnel Stiff grapevine let others know we had arrived. We heard a thump on the door and opened it to find Miles and Pat Moore, wondering if we had the coffee on yet. They lived in a park a couple of miles farther up the canyon from us. It was so good to see them.

The Semlers also lived farther up the canyon, which is where we met J.D. Welty, his wife, Marge, and their family. J.D., a big man, looked like a long, tall Texan. Marge was slim with red hair. Their children were in the same age bracket as Pat.

When we worked on the Oroville Dam project, we lived in a trailer court in the Feather River Canyon, which required us to make a long, scary drive on a road without guardrails. Snow fell on the trailer court, above. Our trailer is at left.

It was such a long, dangerous drive to Oroville that we didn't go down unless we needed something badly. Miles' hair had grown so long he asked Jack to whack it off for him. Jack did just that, but the nice job looked professional.

Jack was one of the unlucky Stiffs injured on this job and transported to the hospital. An hour or so after he left for work, he walked through the door with his arm in a sling and his hand taped to his shoulder. He stopped by to tell me he had hurt the end of his finger and was going to the hospital but not to worry about him.

An ambulance transported Jack, and I drove down that dreaded canyon road to the hospital in Oroville. I don't remember how long it took me, but Jack was just leaving the operating room when I arrived. They had to amputate the first joint of his index finger on his left hand. Jack had placed his hand on some rebar when a rock fell and cut through his glove, slicing off the tip of his finger. The clean cut didn't leave enough skin to mend it, so doctors had to take more of his finger off.

When Jack was awakening from the anesthetic, he was madder than I had ever seen him, with some choice words for the surgeon who took off so much of his finger. He spent several days in the hospital.

On his second day, he looked on a stand by his bed and found a plaque from Miles acknowledging Jack for cutting his hair.

To cheer up Jack after his accident, Miles brought him a homemade plaque commemorating the first haircut Jack gave him in 1955.

Supervisors on Gates & Fox Job

SOME OF THE supervisors on the Gates & Fox tunnel job near Oroville, Calif., were photographed recently by a PB&E representative. Top—(l. to r.): Norwood Driscol, master mechanic; Ed Hill, tunnel superintendent; and Earl M. Woodward, project engineer. Bottom—Jack Sweeney, Clay Paulson, and George Boone, walkers; and Bob Richardson, foreman.

The Oroville, California, newspaper featured some of the Gates & Fox supervisors and workers, including Jack Sweeney, seen at lower left.

Jack got a kick out of this plaque. It is one of many jokes that we pulled on each other. That was our entertainment when we were on other jobs too.

Jack returned to work after about two weeks. His whole hand ached with a pulsing pain. I knit "finger blankies" for his injured digit to keep it warm and protect it from bumping something.

It took several years before Jack recovered from the feelings of loss.

This is about all I remember of the Feather River project. We were happy to head for the job in Paradise, California.

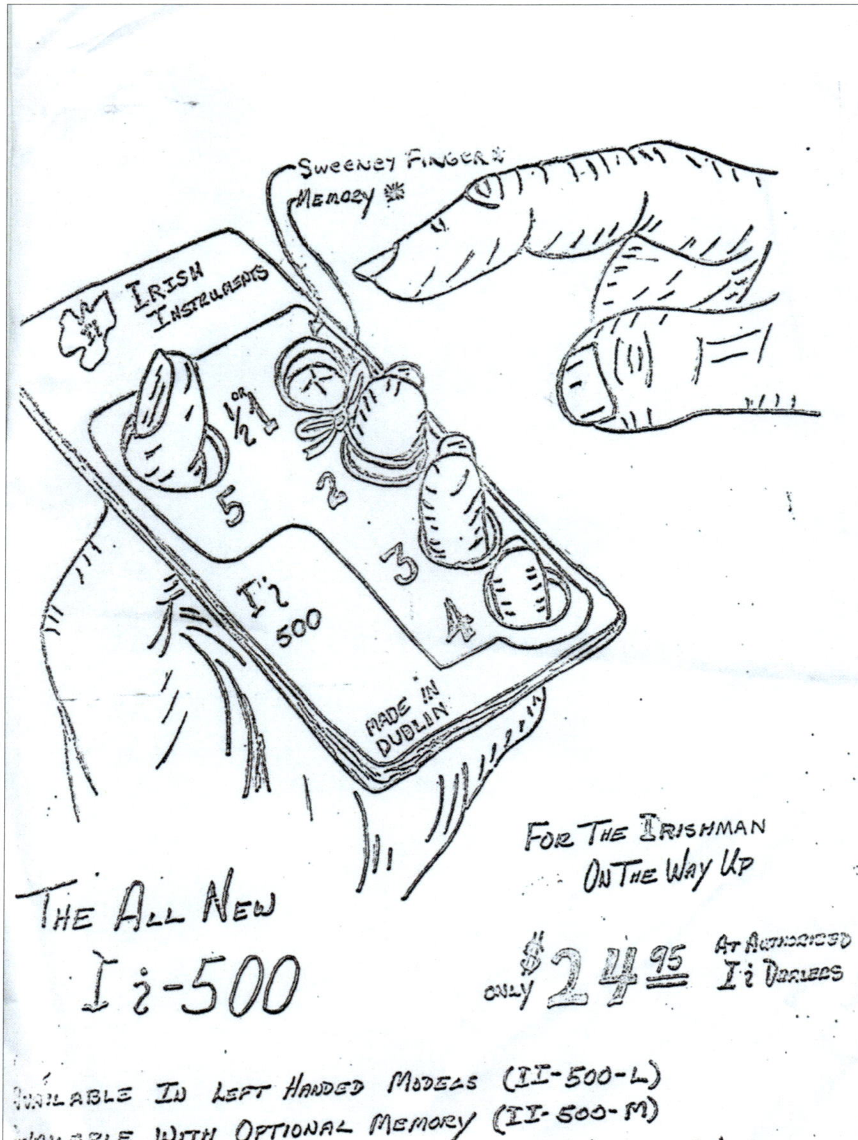

This is someone's idea of the Sweeney Computer, after Jack lost his finger.

Chapter xiii: Paradise, California, 1960–61

North Fork Project

Jack worked on several projects related to construction of the Oroville Dam, which is seen above in this photo from the California Department of Water Resources.

I DON'T HAVE MANY MEMORIES of moving to Paradise, California, where Jack worked for Gates and Fox Construction on the North Fork Project, a Feather River diversion tunnel.

As we alighted in Paradise, we found several places to park mobile homes. We decided on a large park on the outer edge of Paradise with a lot of sand and not many trees.

Mr. O'Malley and Raggs, two Dandie Dinmont Terriers, were brothers who looked a bit like rag mops.

We found a nice spot, opened our trailer, and set up, which was quite a job.

Many different construction companies worked on the North Fork Feather River diversion tunnel. Gates & Fox provided just a small part of the work force.

Kirk Fox dropped by in the wee hours of Christmas Day, plopped a package in the door, and said, "Here's your present."

Jack and I didn't need to open the package to realize one angry little animal was chewing his way out. The mass of moving fur was a Dandie Dinmont Terrier. I fell in love with him immediately and wanted to give him a fancy name. Jack announced, "That's Raggs." He did look like a rag mop. The name stuck. He was a brother to Mr. O'Malley, the dog that the Richardsons received.

When the park owners, Mr. and Mrs. Million, saw Raggs, they said we had to get rid of the animal as they didn't allow dogs, which was different from what they said when we moved in, as we already owned a silver grey cat named Puff (she looked just like a powder-puff). Mr. Million, puffed up with owning a mobile home park, thought he could dictate to the inhabitants.

It took so much work opening and closing our trailer that he didn't think that we would move, but if they didn't allow Raggs, we would. I found a real estate company on Skyway, the main street in Paradise, which just happened to have a nice cement house with a garage for rent on Skyway and Billie Road. The owner said we could rent the spot and find a tenant for the house ourselves, which sounded great to us. The next day, we folded our home and moved to this new address, parking under tall cedar trees with a huge front yard covered with grass. A beauty parlor sat in front of us near the road with a doctor's office directly across the street. Perfect!

The Richardsons parked across town on Case Road, while the Shoemakers and Driscols parked at the Pentz Road Park farther out of Paradise, home to many trailer parks.

On this job we met Art Thoms, his wife, Irna, and their three kids.

By then, Pat was a young man of sixteen. Jack bought a slightly used scooter, painted it Candy Apple red, added a plastic shield to keep the bugs from hitting him in the face, and gave it to him. He was one happy boy.

Jack joined the BPOE Elks Club in Paradise with several of his friends. At their initiation dinner, they ate Mountain Oysters. Jack said they didn't taste bad, but if he had known what they were, he never would have touched them.

His Elks membership added another group of friends to our growing family. Gene and his wife (I think her name was Jeannie) went boating with us. They invited me to try out their new boat. When I did, we had some bad luck. The steering wheel to the rudder broke while we were heading full speed at thirty miles an hour toward the rock cliffs. I was reaching for a life jacket when the boat hit. When I woke up with the

Pat and his scooter buddies join the parade on Main Street in Paradise.

engine cover on top of me, Pat was driving past in our boat. Gene flagged him down, asking him to find Jack and help us out of the water as the boat was sinking. I was scared to death as I didn't swim very well. Jack towed us to shore and I was the only one hurt. My leg was coal black from my hip to my knee, and I suffered a few other bruises. I stood in the cold water to help pull out the boats. I think that saved me from a worse contusion.

I saw a doctor the next day: I suffered only from a bruised leg and ego. That was the last time I rode in a boat, except from shore to our picnic area, where I cared for the little kids while the older ones water-skied.

We went to the Elks' Sweetheart Ball in February. The local undertaker (I forget his name) asked Jack how tall he was. Jack responded, "None of your damn business." I guess he thought he was measuring him for a casket. We enjoyed a good laugh over that one.

Chapter xiv: Madras, Oregon, 1961-62

Round Butte Dam

We photographed a parade on the main street in Madras, Oregon.

JACK WORKED AS A WALKER BOSS for Gates & Fox Construction Company on the Warm Springs Indian Reservation, excavating a diversion tunnel and fifty-two-degree incline shaft on the Round Butte Dam.

Even before arriving in the village of Madras, Oregon, I could smell the tantalizing odor of fresh mint. Madras in the early 1960s was a much bigger town than we had expected.

I headed out from Paradise, California, with our new dog, Raggs, and "Topper," my parakeet, a character. He could say fifty-two phrases, which amazed our friends who often visited us just to hear him talk.

Raggs, another character, was smart as a whip and claimed anything that hit the floor, defending his property. He and Topper got along great. (Topper never landed on the floor.)

Jack and Pat stayed in Paradise to fold our mobile home so the company could haul it to Oregon.

Jack and Pat hooked his new boat to his pickup truck and headed for the job site at Round Butte Dam, on the Warm Springs Indian Reservation.

Everyone was to check in at the Wests' motel on the outskirts of town. George and Joanne "Johnnie" West were proprietors of the good-sized motel with sixteen new units recently added.

Thankfully, they loved animals, especially dogs, and Raggs made quite a hit with them. They had several dogs themselves.

George West, seen on the Plateau behind the motel with Madras in the background, is holding Molly, a seven-pound poodle, while his eighty-pound English sheepdog, Peana, is at his feet.

The rest of the company—the Richardsons, Hills, Shoemakers, Knutsons, and Driscols—arrived by dribs and drabs, checked in at the motel, and started looking for places to live.

Madras boasted several mobile home parks and motels. We decided on a site on a mesa overlooking the town closer to the job site, in the large Plateau Mobile Home Park. Although still surrounded by sand with no grass in sight, large sagebrush bushes, some as big as small trees, helped in the shade department.

I think the main product of Madras at that time was mint, grown in fields with large open culverts lined with cement for irrigation. The air smelled so wonderful—when the wind blew just right, a tinge of mint wafted on the breeze. I think the dogs and kids took advantage of the irrigation ditches on hot days.

Madras had quite a few businesses, among them a Safeway store, bank, bowling alley, golf club, and the Elks Club, in addition to the usual diners and restaurants. Sonny's, a beautiful restaurant and lounge on the southern edge of town, proved wonderful for dining and social meetings.

The Warm Springs Reservoir offered recreation for picnickers and boaters. Jack was in his element now. But he had sold his Punkin' Seed hydro boat in Paradise, so he finagled a great deal on a used hydro-racing boat. He found time to play as well as work. Believe me, all work and no play wasn't Jack's bag. He was never a dull fellow!

Quite a few of the Warm Springs Indians worked on the Round Butte Dam job, which required excavating a diversion tunnel and fifty-two degree incline shaft. Workers used steel tunnel supports and rock-bolts in the tunnel, and steel supports and rock-bolts with wire mesh in the shaft. Not many Native Americans worked in the town, where they did their shopping, visited the beauty parlor, and conducted other business, but mostly kept to themselves.

I was sitting around with too much time on my hands and unaccustomed to having nothing to do. I was still oil painting, but wanted to get out and really do something. I jumped at the chance when Johnnie West hired me to help with cleaning and all around work in the motel. I didn't realize how hard that work was until I filled in every other day to give Johnnie some free time.

We met Bud and Millie Saunders and their family, from Redmond, Oregon, on this job. They were also boaters, owning a vessel with a galley and a changing room downstairs. We often traveled to Haystack Reservoir just outside of Madras to boat. Many of the Gates & Fox "family" joined us for a picnic and boating.

We found our favorite dining place for Chinese food in Prineville, Oregon, which George and Johnnie loved. I must admit it was the best Oriental food I had tasted.

Tunnel Stiffs went by all kind of "handles" or nicknames, with "Loading Pole" and "Haywire" a couple of the oddest ones. Loading Pole was rustling a job in Madras when he and another fellow whom Jack knew quite well from other jobs stopped overnight at our place while they checked out the work site. Jack didn't even know Loading Pole's real name.

Years later, when I returned to Oregon, I remember seeing followers of the Bhagwan Shree Rajneesh riding around in gold Cadillacs. The Bhagwan Shree Rajneesh religious sect bought a 65,000-acre ranch in the Antelope Valley, a few miles from Madras, wishing to build a large ashram, or religious retreat. We heard local gossip contending the Rajneeshees wanted to take over Madras, and in September 1984, they amended Antelope's charter to change the community's name to Rajneesh. Eventually Rajneesh was indicted on immigration charges and later returned to India. In November 1985, the community switched its name back to Antelope and the former ranch is now owned by Young Life, a nondenominational Christian ministry reaching out to adolescents.

Jack sold the boat he built, above, to another Tunnel Stiff, Earl Woodward, and bought a used boat from a lawyer in Los Angeles. Below, Jack and I are seen with that new boat and our dog, Raggs, at the Round Butte recreational area in 1962.

The Round Butte Dam on the Deschutes River in central Oregon is seen in this photo from the National Oceanic and Atmospheric Administration.

Chapter xv: Monroe, Washington, 1962-63

Sultan Dam

Jack was a walker working for Gates and Fox Construction Company on the Sultan Dam diversion. Workers used equipment to move dirt and create the diversion tunnel.

We followed Gates & Fox Construction Company to a job on the diversion for the Sultan Dam near Sultan, Washington. Jack worked as a walker on this job, which required the use of high-tensile strength rock bolts on the abutments and some steel supports and rock-bolts in the tunnel and extensive rock-bolts around the outlet portal.

Arriving in nearby Monroe, we stayed in a motel until we could decide where we wanted to park our mobile home when it arrived. We looked at the fairgrounds and a mobile home park near the motel in Monroe, but they weren't what we wanted. We drove to Sultan and didn't care for the park there either. I asked around Monroe about parking our mobile home. The man who owned the old Standard Oil repair barn on Ann Street told us to look at that and see if it fit our needs.

We liked the place as soon as we saw the huge garage with an upstairs, which Standard Oil had used for the office. It proved a perfect place for Pat to work on his car, and was not far from the junkyard where Pat pulled his wagon to find auto parts he needed. Also, the upstairs could be used as a place to do my oil paintings. When our mobile home arrived, this is where we put down our roots.

We parked our trailer in an industrial lot on Ann Street in Monroe, Washington. It wasn't the most beautiful place, but it served our needs. Pat's 1956 Studebaker Hawk, right, and our Oldsmobile Starfire are parked on the street.

The city of Monroe was beautiful. The Skykomish River ran along the edge of town, with a large bridge across it, separating Monroe from the highway leading to Sultan.

We found everything we needed in town, where we met congenial people. It didn't take long to make friends.

The Driscols parked in Sultan, while the Richardsons found a nice spot in Monroe, within walking distance from us. I think the rest of our group settled in Sultan to be closer to the job.

Pat landed in trouble the first day he drove into the city of Monroe. We received a call from him at the sheriff's office, where he was nabbed for having a "noisy muffler" on his car. I think they were just waiting for someone to arrest. After we saw the sheriff, they gave Pat a warning to fix his muffler. He did.

Kirk Fox, who was born and raised in Bellingham, Washington, had several relatives in Monroe. His cousins, Elsie and Gordon Tjerne, owned and operated the local drugstore. Gordon was the pharmacist. His aunt, Olive Bristol, Elsie's mother, lived in Bothell, Washington. (She was an artist who visited many countries around the world, studying art).

Elsie would often stop for a visit on her daily walks when she was in our area. They also joined our circle of "family" friends and visited us on other construction projects. After Olive heard that Tillie and I liked to paint, she drove the sixteen miles to Monroe from Bothell every other week. We packed lunches and Coca-colas and drove around until she saw a likely place to paint a landscape. We spent the day painting. On rainy days we climbed up to my "gallery" in the garage to paint.

When Tillie and I drove to Olive's in Bothell a couple of times a month, she not only gave us painting lessons but prepared a luncheon made from the different countries in Europe where she had studied. She was eighty-three years old at the time. We learned so much from her. I considered her my "first" art teacher.

While working on the Sultan Dam, a hurricane-force windstorm blew through the area, lasting several days. The Columbus Day Storm, described by the National Weather Service as the top weather event of the 20th century in Washington state, toppled 15 billion board feet of timber along the West Coast from California north to British Columbia. The miners, wielding gas chainsaws, cut up and removed trees that toppled across the roads, blocking access to the job site. The storm was really bad the second day, with winds so strong it blew trees down from across the river. Jack and the other miners huddled behind big equipment to escape the falling trees. One of the workers dove underneath a big heavy truck. A tree fell across the river, struck the truck, and smashed it like a toy, killing that miner immediately. I think his name was Jack Wilson. This was the only fatality on this job.

Jack Wilson had a parrot pet, and spent a lot of time teaching it to say things. The parrot spoke in a voice identical to Jack's. His wife found a new home for the bird after her husband died. Every time it would talk, Jack was there. I know he left a wife and a couple of kids. They were good friends of the Driscols. It really hit home.

Pat graduated from high school while we lived in Monroe. Many of our local friends and most of the construction crew joined us for his graduation. He and his graduating friends held a party. I don't know what time he returned home, but I heard him in the bathroom retching up his toenails. When I asked if he was sick, he mumbled, "Uh-huh." I asked if he had been drinking. Again, I heard a mumbled, "Uh-huh." He said it was awful going down and worse coming back up. I don't believe that he has been very fond of drinking since then.

Pat found a job as a "chicken catcher." A company that raised chickens for marketing employed him and several of his buddies. They waited until late in the evening, after the chickens settled down for the night, then ventured into pens in the dark with flashlights and grabbed the roosting flock, thrusting them into wire cages to be shipped for processing. He would return home in the early morning hours with orders to "undress outside and hit the shower" before he traipsed through the house. Any chickens hurt in the catching were sent home with the kids, as the processors wouldn't accept hurt or lame birds.

I wasn't about to go through the scalding, plucking the feathers, and butchering of the birds, but Tillie Richardson didn't mind. They enjoyed several chicken dinners.

A waitress complained about the kids stopping in for coffee on their way home. She said not only were they covered in feathers but they had a foul (no pun intended) odor about them too.

Jack enjoyed his boat immensely while we lived in Monroe. The man we rented from invited us to his home on Lake Washington, during the famous Seafair hydroplane races. We could sit on the deck of his huge lodge to watch the race live while seeing it on television at the same time. It was wonderful.

Roy and Mary Knutson and Jack and I attended the 1962 World's Fair in Seattle. Jack is seen later with his mother enjoying dinner in the restaurant of the Space Needle.

One weekend Jack participated in a hydro race on Lake Sammamish, accompanied by a rooting squad made up of the Richardsons and Shoemakers. He won a trophy for traveling seventy-one miles an hour in the shortest time.

I don't remember if we had to buy tickets for the 1962 World's Fair in Seattle, but what a wonder it was! We rode the glass elevator all the way up to the top of the "Space Needle" (under protest on my part, as I was scared stiff) and rode on the Monorail. It was all a great wonder to us.

After spending the day walking around the World's Fair and seeing all we could, we took off for home in Monroe. We decided to stop at an Elks Lodge along the way for refreshments, but they wouldn't let Mary and me in because we were wearing slacks. That mode of dress wasn't allowed in the Elks and many other establishments in the '60s. They considered it bad taste.

New wider highways built to accommodate the fair bypassed many of the smaller towns but provided a much shorter trip to Seattle. In fact, during the 1960s, several highways across Washington state were widened into freeways without tolls. The face of the nation was changing.

Chapter xvi: Colfax, California, 1963-64

Rollins Dam

The Rollins Dam tunnels are seen at the beginning of the project.

IN 1963, JACK WORKED for Gates & Fox Construction Company as a walker boss on the excavation and concrete lining of a tunnel at Rollins Dam near Colfax, California.

I don't remember much about this job. It seems like it was of short duration. The Richardsons, tired of trailer living, bought a home in Colfax and we parked our mobile home in their yard and rented a hunting cabin in the hills above Colfax, so we could give the trailer a good cleaning, inside and out as we were making plans to buy a new trailer. We lived in the cabin when Tillie called me one morning and asked if I had the television or radio on. I didn't. She told me that President Jack Kennedy had been shot in a motorcade in Dallas, Texas.

When he died November 22, 1963, Tillie's dad, Alen Downey, who was a politician in Glendo, Wyoming, suffered a heart attack and wound up in the hospital. Tillie planned a big Thanksgiving dinner and invited Kirk Fox, Earl Woodward, and several others to spend the day with them. I told her to go home to her dad.

Jack and I locked up the cabin and moved into the Richardsons' house, bag and baggage. The assassination of President Kennedy consumed television coverage; we didn't even see any advertisements until after his funeral.

I cooked the dinner and, while dining, with our eyes glued to the television, we watched as the authorities brought Lee Harvey Oswald, the accused assassin, into the courthouse. As they pulled him around, we watched, horrified, as Jack Ruby shot Oswald at point blank range. He died shortly afterward.

Lyndon Baines Johnson was sworn in as United States president while on the flight taking JFK's body back to Washington, D.C.

Jackie Kennedy witnessed the swearing-in, still in her bloody suit from the assassination. We stayed at the Richardsons' home and took care of their two boys and the animals for a couple of weeks until they returned from Wyoming. A short while later, the company finished the Colfax job and we moved on to Placerville, California.

The completed Rollins Dam tunnels are seen in this photo, used with permission from Gates & Fox Construction Company.

Kirk Fox, president of the Gates & Fox Construction Company, wrote his memoirs before he passed away. This photo, taken at Georgia's Buford Dam in 1952, graced the cover of his book. Fox said he met Clay Gates in Wyoming in 1947. Two years later, they formed the Gates & Fox Construction Company.

Chapter xvii: Placerville, California, 1964-65

White Rock Tunnel

I am sitting in the Knutsons' flower garden with Mary Knutson and Jack's dad, Neil Sweeney.

OUR NEXT JOB TOOK US back to California, where Jack worked as a shift boss for Gates & Fox Construction Company, excavating the White Rock Tunnel on the American River near Placerville, California.

After working on the Rollins Dam, Gates & Fox Construction moved to the White Rock Tunnel project, excavating about two-and-a-half miles of tunnel and two shafts—one forty-two feet in diameter and the other

July 26, 1964

BOOM! Workmen prepare for another round of blasting, above, in the Sacramento Municipal Utility District's White Rock Tunnel near Placerville, El Dorado County. It will channel water into the White Rock Power Plant on which construction will begin next year. At left, tunnel inspector Frank (Smitty) Landers and worker Pat Sweeney walk toward the face of the tunnel, which will be 23,000 feet long and 24 feet in diameter.

Pat Sweeney tried his hand as a Tunnel Stiff while attending Sierra College in Auburn, California. The job helped with his college expenses in the fall. *The Sacramento Bee* granted permission to use this July 26, 1964, clipping.

thirty-two feet. The company used very few steel supports in the tunnel, but employed both rock bolts and wire mesh in building the shafts. Jack worked as a shift boss on the job.

We traveled to Placerville, California, an old mining town at the center of the California Gold Rush after men building a sawmill for John Sutter in January 1848 first discovered gold in Coloma. Lots of rich history permeated this area.

We found a nice mobile home park on Cold Springs Road close to Placerville and not far from Colfax and Auburn, if you took the back way or shortcut across the American River. We had a time moving in and getting settled here.

Although the Richardsons bought a home in Colfax, Bob Richardson drove the back way to work on the White Rock Tunnel every day and Pat lived at home and attended college in Auburn.

The Knutsons also moved into this big park with large spaces. Most of the inhabitants tended nice lawns and flowers. The others were mostly Tunnel Stiffs we hadn't met yet.

We made the mistake of attending a motor home show in Modesto, where we saw fabulous homes made especially for the show with pneumatic pumps on the hitch that opened and closed the homes. We drooled over one especially, a home that had everything—fine, white French provincial furniture, coffee and end tables with grayish-pink Italian marble tops, bedding, pictures on the walls, flowers, lamps and even a crystal chandelier from Austria in the dining room. All a person needed was cooking utensils, dishes, and clothes to move in. Nah! We couldn't afford that.

We returned the next weekend to look at it again. The dealer said he would make us a good deal on it, as the show was closing the following week. Back at Placerville, we did some serious thinking and decided we could afford the $19,000 price he offered. Although it cost lots of money, we earned good wages on this job and the company would move us when the time arrived. We had never lost any money on the four mobile homes we bought. We received as much, if not more, as a trade-in on a newer, bigger mobile home.

We bought it! In the meantime, the old trailer had to be moved. It remained in perfect condition. One of the walkers on the Gates & Fox job looked at it and decided it was just what he and his family wanted. Done deal? No way. The family was black, and the park owners didn't allow colored people to live there as prejudice still reigned. The family moved all the way to Rawlins Trailer Park and drove back and forth to the job. I don't know who bought the Roadliner or where it was moved.

We decided to park our new rig on a large lot on a knoll above the park, which gave us a big yard with lots of trees and grass.

Jack decided he wanted to work for the Bechtel Corporation, the largest construction company in the world. He figured it offered him more of a chance of moving up in the mining business. He applied for a job and was hired to work in Leadville, Colorado. Jack's parents, visiting from Pennsylvania, accompanied us on our new venture.

Pat was working in the mountains, quite a ways from Placerville, as a forest ranger. I tried to contact him to let him know that we were moving. I left a message with his supervisor to tell him. He never received the message, so what a shock he experienced upon arriving home for a weekend to find not only his parents gone, but his home too. About this time, Pat met Nancy Dunlap, his future wife, in Lotus, California.

Gates & Fox and our "family" group gave us a farewell party in Coloma at the Vineyard House, a very old kind of spooky-looking place that had been there

George and Johnnie West came for a visit from Madras, Oregon, and brought their dogs. Pat is seen at left in his little red 1959 Fiat with Raggs and the Wests' dog.

I am standing at the sliding glass door of our new mobile home, and Jack is seen after opening it to its full width. At right is our dining room and below is our kitchen.

I loved our new home—so did Raggs. The photos above show our living room. At left is Pat's bedroom with dresser, and below is our bedroom, part of the bathroom, the walk-in closet where I kept my sewing machine, and our dresser.

since the stage used to pass through that area. Stage drivers used to put any prisoners they were transporting to jail in an old wine cellar behind the hotel for safekeeping while they rested for a day or so.

The restaurant in a three-story building provided wonderful food and service. The basement remained empty, except for an old-time piano. We asked the owners if we could go down there and dance. Jack's mother was a "honkey-tonk" piano player from the days when she was young and played for the silent movies. We kept her thumping the piano while we all danced until the wee hours. The company gave us a silver chafing dish as a going-away gift.

Jack and I, along with Jack's parents, stayed in a motel that night so the hauling truck could hook up to our "house" for the long trek to Leadville, Colorado. When our mobile home was closed up, it looked like a huge square gray elephant behind the truck.

Jack's folks saw parts of the West on our way to Colorado. We stopped at places of interest, and they loved it.

This painting depicts my impression of the old wine cellar in Coloma, California.

Above, the old James W. Marshall cabin in Coloma, California, posted to Wikimedia Creative Commons. Marshall discovered gold near Coloma on January 24, 1848, which started the Gold Rush. At left is a photo on the historic James Marshall monument in Coloma, courtesy of the California State Parks Department.

Reno, Nevada, was a small town, with Sparks just a little village abutting it. I think we saw only a couple of casinos, Harrah's and the Freemont Hotel.

We heard a rumor that Frank Sinatra was thinking of building a night club and casino in Reno or Las Vegas.

We spent the night at the Freemont, so the folks could do a little gambling and take in a show. The theater, although nice, wasn't much bigger than the theaters where we watched motion pictures.

I went to the foyer to sign up for the night's show, and who came ambling through but Ed Ames in his blue jeans and jacket. He had just walked in to do the show. They didn't have bodyguards or entourages. He just drove his car into the parking lot and entered the hotel to change and put on his show. I did get to say a few words with him. I was used to seeing him in the movie where he played "Mingo" and that is how I greeted him (I felt so embarrassed).

What a loud, strong voice he had! He sang "Dream the Impossible Dream" which gave me goose-bumps.

He really didn't need a microphone. Patti Page was a tiny woman, which surprised me. When she sang on the radio or records, her voice boomed. She was Jack's favorite female singer. A very young Wayne Newton, also on the billing that night, still looked like a child.

Waiters walked through the theater during the performances taking orders for drinks and snacks.

In the morning we stopped at Harrah's and played the slot machines for a while. The room was filled with other gambling machines, but we preferred the slots. The machines took and paid out real coins—from dollars all the way down to pennies—so they could drain the last cent out of a gambler. I liked the whirr of the slots when I pulled the lever and the clink of coins when they paid out. I decided to splurge and put a few silver dollars in the Big Bertha slot machine. Bells and whistles resonating from the huge machine could be heard on the street when somebody hit a jackpot. I hit the hundred dollar jackpot right off the bat! As I scooped up the money, I dropped another coin in the slot and hit another jackpot. I did that four times in a row. My in-laws grew so excited they resembled chickens picking up corn, scooping up the coins that hit the floor as they bounced into the payoff tray.

Waitresses brought free drinks to anyone playing the machines, but I was too busy scooping up my loot.

The dinners were fabulous, not expensive at all. We enjoyed a wonderful time. Jack's dad wanted to see the

Above, Jack is seen with his parents, Ann and Neil Sweeney, in Reno in 1965. In the photo below, I am with Jack's parents outside The Old Original Bucket of Blood in Virginia City in 1965.

Danny Thomas show at Harrah's, but it was booked. He was disappointed, but I think seeing the Bucket of Blood Saloon made up for it. A person could drive as fast as the car would go in Nevada, which had no speed limits. I found out (with white knuckles and Jack driving) that our car could travel 120 miles an hour.

It was a great trip and all the Sweeneys enjoyed this adventure en route to Leadville, Colorado.

Chapter xviii: Leadville, Colorado, August 1965–January 1967

The Homestake Project

This photo, used courtesy of Big Earth Publications in Boulder, Colorado, shows Leadville in the late 1800s with Mount Massive in the background.

JACK WORKED FOR BECHTEL CORPORATION as a chief tunnel inspector on the Homestake Project in Leadville, Colorado, the highest incorporated city in the United States, lying two miles above sea level and almost above the tree line. The altitude is so high some people have trouble adapting.

I don't know how to begin my story on Leadville and surrounding area, where we found so much to see and do. Jack's folks accompanied us from Placerville, California. I'm sure they had a lot to talk about when they returned home to Pennsylvania.

We stopped in Denver and spent the night there, doing some snooping before our long, breathtaking drive to Leadville. All of us were tired when we arrived. We stayed at the Silver King, a very beautiful motel with a restaurant that specialized in steak dinners and lots of other mouth-watering dishes. We enjoyed the meals as it would be a week or more before our mobile home arrived.

In the meantime, we checked out the mobile parks and decided on one in the pines and firs. Aaaah! What refreshing air and atmosphere!

Jack's folks couldn't stand the altitude and left for home after just a few days at the motel. They did see a lot of Leadville before leaving. As we accompanied them to the Denver train station to go back to Pennsylvania, we snapped photographs of them throwing snowballs at the Continental Divide on August 9, 1965.

Ann and Neil Sweeney are seen at Loveland Pass as they were leaving Leadville on August 9, 1965.

Our mobile home finally arrived! After the hauling company set it up in the park, we were happy to move back in. We encountered no problems making acquaintances. The first day, after Jack left for work, I heard a knock at my door. A few of the women whose trailer homes were across the road from ours, burst into my home uninvited to "kidnap" me and take me to a "Come as you are" breakfast. I was showering and, if it hadn't been for Raggs chasing them off, I would have been out and about in my "birthday suit," dentures still soaking, my hair a soggy mess. They let me change, then led me off to breakfast in town with quite a few other half-dressed women. That was my welcome to Leadville. I knew this was going to be a fun group.

Some of these women were "dump diggers," who scavenged in places that people from Leadville and surrounding areas used many years ago during the mining heyday, uncovering many old treasures—bottles, tin cans, whatever other antiques they discovered. They dug so deep they used ladders.... I was fascinated with their findings, but never brave enough to join them. They sold a lot of their treasures to antique dealers.

The fascination of Leadville lay in its history and old buildings. A house of prostitution remained active in a saloon/restaurant (I think it was the Pioneer Saloon). The "ladies," dressed in western apparel, served as barmaids. The beautiful original old bar took up most of one wall, along with a piano shipped over mountain passes from San Francisco. The saloon also featured old mirrors, brass spittoons, and a piano player tickling the ivories—just like in a western movie.

Leadville, home of many self-made millionaires in its heyday, including the Silver Kings—I loved everything about the community.

The high sidewalks on Chestnut and Harrison streets, with cars parked alongside them, amazed me as one had to climb several steps to reach the sidewalk, houses, and shops. I read in a local book that the sidewalks were built

Baby Doe Tabor is seen outside the cabin of the Matchless Mine, started by her husband, Horace Tabor. The silver mine is in the center photo. At left is the House with the Eye in Leadville. The photos are courtesy of Big Earth Publications in Boulder, Colorado.

that way in the old days when streets were not paved, just thick mud, and everyone traveled by horseback or horse-pulled buggies or carriages.

These sidewalks were hard on car doors in this modern age; many inhabitants of Leadville sported scratched and dented car doors.

Staff Sergeant Barry Sadler, who wrote "The Ballad of the Green Berets," a native of Carlsbad, New Mexico, grew up in Leadville in his younger years.

The back road to Aspen, Colorado, was a narrow, dirt road, alive with quaking green and yellow aspens. I felt as though, if I listened hard enough, I could hear the leaves tinkling a tune. We loved the beautiful drive to Aspen, which at that time featured just one ski lodge and a single ski lift. Jack swapped boating for skiing on this job. I tried skiing a bit, but spent most of my time trying to dig myself out of snow banks. Outdoor life wasn't for me. I accompanied him and spent my time by the huge fireplace in the lodge, sipping a hot drink of coffee, watching the skiers.

Many times, after seeing the skiers leave for the slopes, I stayed home and prepared a hot dinner for everyone—usually a big batch of spaghetti, salad, and garlic bread. Few of them felt any pain as they sipped from wine flasks slung over their hips, which proved a very important part of their ski apparel.

I obtained the recipe for "glog," a wine concocted from fruit that took a long time to make, but they loved it. I served the beverage hot, in mugs, a very welcome treat on cold days.

Vail Ski Resort, today the second largest single ski mountain in North America behind British Columbia's Whistler/Blackcomb, opened in December 1962 with only one small ski slope and a shed to stash skis and sip hot coffee.

The actor Jerry Lewis was building a large home in a big grove of trees. People thought he was nuts. If we had only known then what Vail would become in the future, we would have invested in property as it was dirt cheap at that time.

Cooper Hill Ski Area was a short way from where we lived on the outskirts of Leadville. Jack and Don Orcutt kept their skis there and stopped on the way home from work, if it was still daylight, for a couple of trips up and down the slopes.

Don lived in a trailer right behind us in the park. He owned a large dog, I don't remember what breed, but he was big. Don named him Dammit. He was the first dog I ever saw that loved vegetable salad. Don's girlfriend, Lynda, from Farmington, New Mexico, spent a week at a time visiting him. I think she spent more time at our house. She was a beautiful girl; if I could have had a daughter, it would be her. We stayed close for years. Even after they married and moved away, Don and Lynda visited us wherever we were living.

We joined a bowling league. Women played in the morning and the men at night. Although I didn't find a job there, I did keep score for the men's league. Many of the miners from the job played in these leagues. I was a lousy bowler too! I earned a trophy for landing in last place every time.

Mr. Gates, the local owner of the alley, asked me to do some paintings for him, which I did. Thrilled with the paintings, Mr. Gates paid me by giving both Jack and me bowling shoes and brand new bowling balls—a black ball with red racing stripe for Jack, with his name engraved on it, and a marbleized ball for me.

We added Don and Lynda Orcutt, John and Jean Larch, Dick and Jane Perely and their families to our Tunnel Stiff "family" on this job. They are still on my Christmas card list!

A young engineer from Japan, I think his name was Nishi Nouri (at least it sounded like that), came to Leadville to show how a piece of equipment "made in Japan" worked. Jack invited him to dinner at our home. He wasn't allowed to touch alcohol of any kind, or

We stopped and parked our Ford Mustang to take photographs on Independence Pass between Leadville and Aspen, Colorado. Jack, seen at right in the photo below, was a snow bunny with Don Orcutt on the ski slope at Vail, Colorado.

receive gifts or souvenirs from the United States. He asked me about each dish I made —a pork roast, mashed potatoes, and the usual American dinner.

He grew quite incensed when I asked him if his wife drove a car. No way could a woman drive; she stayed home, cared for their baby daughter, and did the housework. He did purchase two hunting rifles in Denver, which he was proud of, explaining that they cost too much to buy in Japan.

We found him to be very interesting and well educated.

Nishi Nouri, a young engineer from Japan, is seen above seated next to Jack at our trailer home in Leadville. I painted the picture hanging above their heads.

We did quite a bit of sightseeing while we were in Colorado. One of the most amazing sights we saw was from the top of Pikes Peak—a view enjoyed only after traveling a long two-way road to the top, with curves so sharp they put mirrors along them so travelers could see any hazards ahead. When we reached the top, we were greeted by huge triple-wide manufactured buildings used as a restaurant, motel, and visitors' center. I couldn't imagine how they transported them to the peak, unless they were flown in. They proved quite a shock to the senses.

It was an awesome drive up the mountain to the peak, but much worse on the harrowing trip back down. It was quite an experience for us.

I am seen at the summit of Pikes Peak.

I met people from an art league who were hosting a show at one of the mountain lodges in Breckenridge, Colorado, and they asked me to join them. I did palette knife paintings at that time. This was the first time I ever participated in an art show. I did earn an honorable mention for the painting of Burro Race over Mosquito Pass.

I did most of my shopping in nearby Climax, Colorado, at a big store that looked like an Army PX, or Post Exchange. I think it mostly served the miners who worked in the local molybdenum mine. In this mine workers drilled from the top of the mountain down inside, which was very odd-looking.

My dad flew in from New York state for a visit. My mom had passed away in 1964.

The altitude bothered him, but he stayed. We introduced him to the fellows at the Leadville Elks Club. Several of the members asked him to drop in any time he wanted to play cards and gab. We appreciated that offer as life could be pretty boring sitting around doing nothing. We let him use our car and noticed a lot more scrapes and dents when he returned home. He also attended our bowling matches, where the folks welcomed him and rubbed his head for good luck. He liked that! Jean Larch picked him up several times a week for rides to show him the beautiful areas around Leadville. Her kids had never had a "grandpa" and they adopted him.

Nancy's brother, Allen Dunlap, decided to drive with Pat from Placerville to Leadville so they could work in the tunnel. We were so happy to have Pat home again; our nest was empty without him, after we moved away.

They worked for the summer. Pat bought another car in Denver just before he and Allen headed back to California. He was antsy to return to Lotus and Nancy. We took my dad to Denver to catch his plane back to New York.

Not long afterward, we received a call from Pat, who said he and Nancy were getting married in Reno, Nevada, on June 14, 1966.

Many of our Gates & Fox Construction "family" visited us—the Richardsons, Ramseys, and "Gordy" and Elsie Tjerne, to name only a few.

Roy and Bertha Shoemaker had retired and moved back to their home in Fort Collins, Colorado. The day they visited we had just heard the news that "Buddy" Fox, Kirk's only son, had drowned while heading to their home from Grants Pass. His car hit a slick spot in the road and plunged into the Rogue River where he was trapped and drowned. The Shoemakers spent the

My dad, Charlie Thomas, visited us in Leadville in early 1966. He is seen above between me and Pat in this March 6 photo. At right, Jack is at the job site in Colorado, and below, I'm at the dining room table in our home.

In September 1966, I participated in my first art show ever in Breckenridge, Colorado, where I earned an honorable mention for the painting of a Burro Race over Mosquito Pass.

night, then drove to Oregon to comfort the Foxes. It was a very, very sad time for all of us.

Toward the end of our job in Leadville, Kirk Fox visited and asked Jack to come back to work for him. He had a job in Crystal City, near San Mateo, California, and decided Jack would be an asset. Jack said he would think on it.

After the Homestead job finished in Leadville, Bechtel moved to another small job—the Rampart Tunnel near Littleton, Colorado. We moved to Aurora, on the outskirts of Denver, and Jack worked as the chief tunnel inspector.

Pat served in the Army Air Force, stationed in Texas. He and Nancy visited us on their way back to California. Traveling cash was always welcome. We dug out our cash stash jug (it really was our "Reno" money—all quarters and silver dollars). It paid for their trip and meals en route to California.

When the Rampart Tunnel job ended, we headed to California for the Gates & Fox job in Crystal Springs.

Chapter xix: San Mateo, California, January 1967-August 1967

Crystal Springs Bypass Tunnel

The Crystal Springs bypass tunnel, built by Gates & Fox Construction Company, is seen in March 1969.

JACK RETURNED TO WORK for Gates and Fox Construction Company in January 1967, when he served as superintendent on the Crystal Springs bypass tunnel near San Mateo, California.

Gates & Fox Construction Company built the Crystal Springs bypass tunnel using the Scott Drag Tooth Mole seen in these company photos, used with permission.

We drove a now familiar route west from the Colorado mountains to the West Coast again. I don't recall how long it took traveling to San Mateo.

We decided to live at Moss Beach, a, huge beautiful park right on the beach, more like a miniature city than a mobile park, with paved streets and sidewalks. We found the Pacific Ocean air a refreshing change from the Colorado mountains.

We discovered our group was scattered around the area. The Richardsons and Woodwards chose to live in the city of San Mateo, while Bud and Millie Saunders settled on Half Moon Bay, a short way from Moss Beach. We met Joe "Red" Nickerson and his family on this job.

Our first night at Moss Beach almost persuaded us to turn around and head off someplace else. A squall almost flooded the park. I watched metal patio covers tear apart and float down our street as the Pacific Ocean really acted up. Everyone picked up their patio furniture and took it and everything else they owned inside. What a welcome! Everything appeared okay with little damage done. Our place remained all in one piece.

I loved the beaches and often took Raggs over the hills to the shore. He ran onto the shore and barked at the waves; when they receded, he thought he had chased them back. Such a character!

I found treasures galore on the beach, such as star fish and all kinds of shells. When the tide receded, Raggs and I walked quite a ways out from the shore to find "goodies." One time we saw a dead sea lion beached on shore, its corpse too heavy to flow back to sea, growing pretty rancid. The authorities burned the carcass, which took several days.

Another day I found a bottle washed ashore with a note inside. When I opened it, instead of a message washed ashore from some faraway land, I discovered a note from a school kid in Carpenteria, about fourteen miles from Moss Beach. I sent the note to the boy, who

The top photo, taken by Robert E. Nylund in May 2007, shows California's Moss Beach. He released the photo into the public domain via Wikipedia. At left I am seen with Tillie and Bob Richardson, who visited us from the city of San Mateo. Tillie and I painted together during her visits. The Pacific Ocean and Moss Beach are in the distance.

must have been disappointed as I didn't hear back from him. Perhaps it had been a school project.

My sixth sense started clicking in as I had that feeling of being watched—just Raggs and me alone in that big expanse of beach and sand dunes. I quit going by myself, taking visiting friends along with us, or Jack would accompany me when he had the time.

We found a soon-to-be favorite restaurant at Moss Beach: Ida's Fish Shack, right on the shore, serving everything fresh from the ocean every day. We enjoyed their specials, featuring a pitcher of beer, green salad, and all the crab you could eat. We liked eating out on Fridays as we loved fresh fish, especially the sweet taste of abalone pried from underwater rocks by deep-sea divers. The restaurant's Friday specials featured fresh-made clam chowder, crusty bread, and a pitcher of beer. Swordfish was another favorite dish.

Although really only a roadside stand, Ida's Fish Shack drew many customers who stopped to buy fresh seafood for use at home.

Half Moon Bay, where Bud and Millie lived, was another favorite place to spend an evening. We enjoyed sitting on benches at long tables in a huge barn-like building, where on weekends we could buy a pitcher of beer, green salad, and all the crab legs one could eat for $3.00. The food kept coming until we were stuffed. It also featured a dance floor, if you could stand the fishy odor of your dance partners. What a good place to spend a Saturday night!

Tillie Richardson visited from the city of San Mateo, hauling along her painting materials. We climbed the cliffs over Moss Beach and Half Moon Bay to paint seascapes and views of the surrounding areas. We hosted visitors on this job, as we always did, people who wanted to see the beautiful Pacific and where we were living.

Miles Moore had retired; he and Pat lived at the top of a mountainside at Pacifica, California, in a beautiful home overlooking the whole region, including San Francisco. I remained a bit wary of living on the side of the mountain, fearing that it could wash into the ocean in the event of a landside.

Pat Moore, who worked as a nurse for the Red Cross in San Francisco, stopped for a visit when she was in our area.

Jack and Bob Richardson at the Richardsons' home in San Mateo, California, during a visit in February 1968, at left, and Tillie and I at right.

We became friends with a few of the people in the park, playing cards and bingo in our spare time. Red and Tweetie Nickersons' little son, Joe, rode his tricycle down the street to visit me. He couldn't talk plainly at the time, but that didn't stop him from talking away, not making any sense. I encouraged him to slow down and pronounce each syllable for me. He soon spoke like a magpie. I enjoyed his almost daily visits.

Jack worked as superintendent building the Crystal Springs bypass tunnel and two shafts, which required steel supports and concrete.

Some miners used "the mole" at this job site. It was quite a piece of machinery! Instead of the miners having to do the exhausting, backbreaking drilling through the mountain, the mole drilled through the rock face. I believe Crystal Springs was the first time it was used.

Gates & Fox company finished ahead of schedule.

Chapter xx: Tehachapi Mountains, California, 1967-68

Los Angeles Aqueduct

Jack worked as superintendent of the concrete tunnel lining on the Los Angeles Aqueduct.

WHILE WORKING FOR GATES & FOX Construction Company on the Los Angeles Aqueduct, Jack was superintendent of the concrete tunnel lining.

The aqueduct, near Interstate 5, follows what is known as the "Grapevine," a long four-lane highway snaking its way from Sacramento up and over the Tehachapi Mountains, transporting water to Los Angeles.

We lived in Lebec, on the very top of the grapevine, and during that time learned about eighteen men trapped in a tunnel near Gorman, California, on December 21, 1967. All were rescued safely, but these stories resonated deeply with men who worked underground.

I don't remember much about this project. It took awhile to find a place to live as our trailer was too big for spaces at Frazier Park, where Joe "Red" and Marie "Tweetie" Nickerson and kids had parked. Ed and Dot Hill and their family lived a few miles from us in a rented house at Lake of the Woods, just a nice ride for visiting. Bud and Millie Saunders lived in Sacramento at the foot of the grapevine.

George and Mary Veatch now lived in Lancaster, California, so we enjoyed visiting with them again.

We found a nice spot on a hill above Frazier Park, with only houses. We did notice an empty lot, tracked down the owners, and asked if we could settle there. He said if we installed a septic tank, hooked up to the water, and put up the electric pole, we could park there, rent-free, with the option to buy the lot if we wanted it.

The mobile home fit perfectly. Some of the crew members helped set it up with a long covered porch alongside. We had a good-sized back and front yard. However, we didn't know at the time that water was scarce; some days we didn't have any at all.

The mountains soared so high we couldn't acquire TV signals. We worked with a neighbor, deciding to share the cost of connecting to the huge antenna at the top of the mountain. It worked fine except another neighbor who lived on the street behind us kept tapping into the line; then we saw only a fuzzy TV screen. We had asked them to share the cost with us

A cement plant was erected near the tunnel where Jack worked on the aqueduct project.

before we hooked up to it, but they said it was too expensive. We had to find where they clipped into our line and undo it sometimes several times a day. One evening they even had the nerve to call the police, to force us to let them on the line. The officer who answered the call said we had to share the antenna with the neighbor, although he wouldn't help pay for it. Jack unhooked the antenna. Period!

Gorman was a small town with the Caravan Restaurant and a motel down the highway south of where we lived. I believe the Caravan served fantastic Indonesian food, but you had to watch the waiters. The minute you put your knife or fork down, they whisked the plate away, proving what great service they provided. It took some explaining that, just because we put the silverware to rest for a second, we were not done with the dish. Although we loved the food, we didn't like the waiters hovering over our shoulders while dining. But it was a great place to entertain guests. In fact, Kirk Fox held our second going-away party there.

Driving into Bakersfield proved hazardous on summer mornings in the dense fog. We'd see clear, blue skies at Lebec, but hit the dense fog when we dropped down into the valley. In the poor visibility, headlights appeared very dim on other cars and trucks and lots of accidents occurred on this stretch of road. The fog lifted entering Bakersfield. Several times on the trip, I'd find a place to turn around and head back home because of the fog.

Partway down the grapevine, we saw a memorial to the Buffalo Soldiers, the black servicemen in the 9th and 10th Cavalry Regiments formed shortly after the Civil War ended

On this job, Jack met Ralph Winner and his wife, Gladys, from Pierre, South Dakota. Ralph worked with Jack several times and the couple joined us at many events, but we lost track of them after this job ended.

The job site at the top of the Grapevine, showing the tunnel, cables, and even an overturned truck.

Jack, right, and Ralph Winner are seen above working on the Los Angeles Aqueduct project. At left is Norwood Driscol, who helped design the Jumbo, which was equipment used on construction project.

At that time, Jack chose to return to work at the Bechtel Corporation. He wasn't growing any younger and decided to use his "brain instead of his brawn" for a change. When Mr. Dickerson of Bechtel phoned, he didn't need to do much arm-twisting to persuade Jack

Gates & Fox workers, including Norwood Driscol, work on building the Jumbo for use in a tunnel construction project. Below is the finished Jumbo.

At left, Pat and I stand outside our trailer home in Lebec, California. The lower left photo shows our friend, Ed Hill, in 1968, and at lower right is Jack celebrating his birthday August 9th.

to rejoin the company. We finished the Los Angeles aqueduct job, then moved to a new position in Sonora, California, where Jack began work on the new Don Pedro Dam May 1, 1968. Bechtel even gave him a new pickup truck.

Chapter xxi: Sonora, California, 1968-71

New Don Pedro Dam

Mayor Joseph Alioto of San Francisco dedicated the New Don Pedro Dam when it was completed in May 1970. Photo used courtesy of Bechtel Corporation.

WE STAYED IN THE CARAVAN INN and motel at Gorman while we removed the awning and dismantled the mobile home in preparation for the movers to haul it to Sonora, California. Jack worked as chief tunnel inspect or for the Bechtel Corporation on the New Don Pedro Dam on the Tuolumne River.

We then lived at Woody's Motel in Sonora while searching for a mobile home park. We decided to settle at Burger's Trailer Park, just between Jamestown and

Sonora. After the lower section of the park filled up, workers built a road around the hill and constructed an upper section for additional plots, which is where we parked. Behind our trailer, we walked along a little narrow lane to the lower section and across the road to a wooded creek. That's where I did my walking exercise and enjoyed views of running water gurgling over the stony creek.

The huge New Don Pedro Dam project employed many workers spread out in different cities, including Chico and Modesto.

Trailers and mobile homes increased in size and beauty, no longer requiring us to hide behind gas stations or live in the toolies. People took pride in their spaces, adorning the area around their homes with fences, flowers, and vines and proudly displaying painted signs bearing their names to let people know who lived where.

It was like coming out of the dark ages. The mobile home parks no longer housed only "tunnel stiffs." Professional people and retirees lived there too.

I could hardly wait to start exploring Sonora, a historic foothills village near the Stanislaus River where the Mi-Wuk Indians lived when Don Gabriel Moraga's expedition first entered the area in 1806.

I walked along narrow streets with cobbled, cracked, and crooked sidewalks—none were level, just paved over hills and dales.

Jamestown, also called "Jimtown" and "Hangtown," still had the original tree standing in the middle of town, where bandits had been strung up, and old train and tracks from bygone days were still in use. Quite a few Hollywood movies were filmed in that area, and the American television comedy Petticoat Junction brought fame to the train station with its water tower during its run from 1963 to 1970.

While we lived near Sonora, Hollywood producers filmed *The Great Bank Robbery*, starring Kim Novak,

Joyce Hill is seen above in the old town of Columbia Park, California, in June 1968. In the photo at left, I am standing on the right next to Dorothy and Tim Fuller and their daughter, April, at the Stage Drivers Retreat. Below, Beth and Scott Hill are seated inside the old town retreat at Columbia Park in June 1968.

Hollywood filmed The Great Train Robbery while we were in Sonora. Kim Novak signed autographs for fans near the trailers. At right, I am seen behind her, next in line for her autograph.

Zero Mostel, and Clint Walker. The actors ate lunches at picnic tables and benches in a field under a big tree. We were lucky enough to see the movie stars living in trailers on the lot. I met Kim Novak, which was quite a thrill.

We met new folks working for the Bechtel Corp—a lot more people than on any other job we had worked before. The ones we met were very nice. We had so much to see and do. Jack was in "hog heaven" as he could resume boating and other recreational activities that he liked to do.

We went boating, water skiing, bowling, and out to dinner at the many restaurants. In the winter Jack skied not far from Sonora. We frequented one restaurant in particular on weekends—Poor Reds, a barbecue rib house in El Dorado on the outskirts of Sonora. The food was so good customers stood outside in line, waiting their turn.

Bechtel Company held outdoor picnics around the area where we could meet and grow acquainted with the other employees. Ron and Phyllis Gleasman, Dave and Lee Thomas, Brig and Kim Trisdale and their families became lifetime members of our Tunnel Stiff "family" on this job.

What a scenic area! Just out of Sonora was the Mi-Wuk Indian village, where we attended their Acorn Festival, spreading blankets on the ground and sharing the weekend event with Indians from all over America who dressed in their full regalia and danced. I remember seeing a performance by Jay Silverheels, the Canadian Mohawk First Nations actor who portrayed the Lone Ranger's sidekick, Tonto, and other famous Native American actors from Hollywood.

It was very beautiful to see, but we didn't like the taste of the "acorn soup."

Also near Sonora we visited Twain-Harte, where Mark Twain wrote "The Celebrated Jumping Frog of Calaveras County" in 1865, retelling a story heard from

We joined the MiWok Indians at their village for the Acorn Festival on September 15, 1968.

hotel bartender Simon Wheeler at Angels Camp. We saw the historic cabin erected in 1922 to commemorate Samuel Clemens' presence in Tuolumne County during the winter of 1864-65.

We also enjoyed our visit to the wonderful historic town of Columbia, California, which retained its old-time flavor.

I think the old hotel was the one used in sections of the TV series *Bonanza* and *Little House on the Prairie*. Kids could ride on the old stage and horses, and the community staged a bank robbery for entertainment. The old "family-style" saloon retained its historic flavor, allowing kids to hop on a bar stool and order a "sarsaparilla."

The town of Murphys featured a fantastic restaurant with an all-you-can-eat weekend special for $3.50. We could look over the pork and beef roasts cooked on huge spits and ask for the cuts we wanted, cooked to order.

We drove a back route to Carson City, Reno, and other cities for sightseeing and gambling.

I found a small art gallery just outside of Sonora, a tiny, rustic shack on a creek bank with room for a small kitchen and a wood stove, with an area to store easels and paintings. The front featured an old sunroom with many windows where artists took lessons. Jan Gorgas, a local lady, owned and operated the gallery.

An art teacher, Nevin Kempthorn, from San Francisco taught lessons. I wanted to take lessons so badly, but they cost far too much. Jan made a deal with me: If I would start the fire, put on the coffee pot, clean and dust the gallery, and prepare things for painting early in the mornings, my lessons would be free. Boy, I snapped up that offer in a heartbeat! I have never regretted it. I received some of the best art instructions of my life.

Nevin was a great teacher. She started painting as a four-year-old and she taught me how to put life in my paintings, such as water and reflections. I learned more

We visited the Angels Camp Museum near Copperopolis, California, in the 1950s and again in the 1960s. The sign at left describes the history of Hornitas. Pat and I are seen below with Lee Riggles, the new owner of Beauty.

in that class than I had ever known. She would take our orders and buy the best paint and brushes from San Francisco. It was fun meeting so many artsy folks.

One of our projects was to think of a "character" we had met or heard about who stuck in our minds over the years. We then painted that character from memory. I chose "Baby Doe Tabor" from Leadville, Colorado, whose unforgettable story remained with me. Her husband Horace Tabor, one of the Silver Kings of Leadville, went broke when the bottom dropped out of the silver market. Before Horace died, he told Baby Doe to "hang on to the Matchless Silver Mine," insisting it would come back again.

She kept the mine, but went broke, living alone during this time. The feisty lady kept a gun and shot at anyone venturing onto her property. They found her frozen to death in the Matchless shack where she stayed, waiting diligently for silver to "boom" again.

"Baby Doe" Tabor and the Matchless mine in Leadville, Colorado. This is one of my best portrait paintings. Below is a newspaper clipping that shows my art teachers and mentors, Nevin Kempthorn and Jan Gorgas.

OIL PAINTING will be demonstrated in a program tomorrow night in Sonora Memorial hall. Artist Nevin Kempthorne (right) discusses the program with Jan Gorgas of Jan's gallery.

The men working in the tunnels formed the Twilight Bowling League in April 1970 as a fun way to relax. We all went along to heckle and cheer them on.

But on the job site, the men faced dangerous hazards and laborious eight- to ten-hour days in the tunnel, where Jack was chief inspector.

When Ron Gleasman, an office engineer, met Jack, he asked him where I was from. Jack told him it was just a little spot on the map near Rome, New York. Nobody would know where it was. But Ron was born and raised in Clinton, New York, near Rome, and his parents had attended the barn dances at Crowfoots that I went to as a teenager. He was also a cousin to my nephew, Dick Gleasman. What a small world!

Ron and Phyllis became the parents of a baby girl, Terri, while we lived in Sonora.

I think that Dave Thomas worked as a surveyor. I'm not sure what Wheeler, Hensley, and Simmons did on the job. I know that we had a lot of fun with them and their families, and they all became part of our Tunnel Stiffs "family."

As work wound down on the New Don Pedro Dam, Bechtel Corporation was bidding on a job in Washington, D.C., that would employ most of the "stiffs" working the Sonora job.

On May 18, 1970, work on the New Don Pedro dam was finished. Quite a feat! On May 22, we attended the dedication by San Francisco Mayor Joseph Alioto. Bechtel Corporation held a big barbecue, which most of our crew attended.

We packed and closed our mobile home for the trip to Washington, D.C. Jack hauled it to the garage in Sonora to have the chassis inspected to make sure it was ready for the long trip east. I should have had a camera ready when we returned from the garage to clean up our space. When Raggs dashed out of the car as usual and bounded up the steps, the expression on

Jack is seen at lower right in this photo with other Tunnel Stiffs on his Twilight Bowling League team. Other members of the team are, standing, from left, Dave Thomas, Ed Hensley, and Ken Wheeler, and crouching at left is Dick Simmons.

his face was priceless. The steps led nowhere. His house was gone.

We stayed in a motel until the truck driver came to haul our "home" to D.C., where our troubles began.

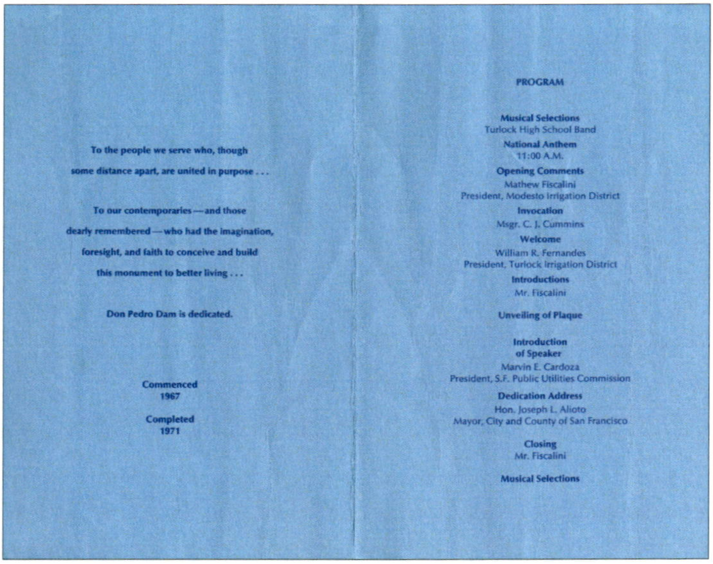

The dedication program for the New Don Pedro Dam, above, used courtesy of Bechtel Corporation.

Chapter xxii: Washington, D.C., June 1971-April 1980

Rapid Transit Project

Jack is seen in the high car in the truck-mounted scissors inside the A 116 Bethesda Station in October 1979.

WE HAD NO DEADLINE to check in on this Washington, D.C., Rapid Transit project, so we decided to visit Tunnel Stiff friends along the way. The Gates & Fox Construction Company built its new headquarters in the Loomis, California, area, where the Foxes, the Hills, the Knutsons, and the Wood-

Jack was chief inspector for Bechtel on the Washington, D.C., Rapid Transit project, which included the K-1 contract with open cut and twin tunnels from Arlington, Virginia, to Rosslynn, Virginia, with one fan shaft; the A-10a and A-10b contracts, with twin tunnels and four shafts at Bethesda, Maryland; and the A-11a and A-11b contracts, with an underground station at Bethesda; and the C-4 Contract, with twin tunnels under the Potomac River.

wards bought land and built homes. The Richardsons were working on the Tennessee Valley Authority project, and the Shoemakers bought land and settled in Flagstaff, Arizona. The Hoilands retired in Oroville, California. We also visited old friends and relatives in Texas and other states en route to Washington, D.C., where a new life awaited us.

We marveled at the inner and outer lanes of the Beltway, also known as Interstate 495, where motorists heading to and from Washington, D.C., Virginia, and Maryland zoomed along at sixty to sixty-five miles an hour.

We had one heck of a time navigating. If we became lost, we just kept driving around the huge circular beltway until we approached an off-ramp to where we wanted to go. We had never seen anything like it! The Beltway looked like a giant ant trail, packed with rushing cars, trucks, and other rigs, all in a hurry, flowing with the constant thrum of tires slapping pavement.

We finally managed to take the off-ramp at Route I-95 and discovered a resting place at the Holiday Inn close to the Beltway. We checked into our room and commenced to settle in for a few days while we found the job site and a place to park our mobile home … when it arrived.

Although the overcast sky looked ominous and cold, we discovered otherwise the minute we stepped out of the car into the hot, heavy, muggy air. Our motel room looked and smelled old and moldy, with an old bed and furniture that looked like it hailed from a thrift shop. Everything seemed damp, as though we could wring the moisture from the bedding. The bathroom proved a disgrace. I don't think the toilet, tub, or shower had undergone a good cleaning in years. A sliding glass door led from our room to a yard in back, where we could take the dog for a walk and play awhile.

My first stop was at a little store nearby for cleaning supplies and disinfectant so I could give the room a good scrubbing. We ate good food in a small restaurant across from the motel, although the area looked worn and ramshackle.

Several times a day, Jack took trial runs back and forth from the motel to Bechtel headquarters in L'Enfant (pronounced L'Fau) Plaza, in D.C., to familiarize himself with how to navigate the Beltway.

Then our troubles began. We called the trucking outfit hauling our home, asking where it was and when it would arrive, only to discover the rig had broken down in St. Louis, Missouri. Flat tires and broken wheels kept it stuck there, with no estimate on how long repairs would take.

It took seventeen days to arrive. What a sorry sight! Boards blocked the front windows so they wouldn't break and road dirt and tar covered the home, leaving us rather deflated.

In the meantime, we had looked for a place to park, but found no mobile home parks anyplace off the freeway. Only after several days of searching did we find a space. We had about given up when we heard of a place near Harwood, Maryland. What a surprise! Maryland Manor Estates, a new park still under construction in a former swampland filled with rocks and sand, was laid out like a city with about 350 large parking spaces. It was near Wayson's Corners on the Patuxent River, a forty-five-minute drive to D.C. and the job site.

According to the park rules, the mobile homes had to be new or not more than seven years old. Ours was much older, but we showed the officials pictures and they agreed to let us in. We put a deposit on the space immediately. The park had paved streets and sidewalks, a huge recreation building, and a swimming pool. As soon as someone rented a space, the park operators lay sod for the lawn. It provided a very beautiful place to put down roots for a few years.

Jack and I are seen in front of the White House while working on the Washington, D.C., Rapid Transit project. My sister and her family visited us and the White House. Below are my sister and her husband, Mary and Charlie Andrews, my nephew, Andy Thomas, and their grandson, Scott Patrick.

As soon as we made our deposit, they started working on our space—still just a lot with sand, trees, and mosquitoes. What a disappointment when our Roadliner home arrived, with the boards nailed across the front, dents in the side, covered with road tar and dust, both sewer and water lines broken. Good thing we put our deposit down before it arrived.

Jack is seen above working on the K-1 Station in August 1979. The station at Bethesda is seen that same month in the photo below.

Before we packed our trailer to move it, I gave it a good cleaning, putting trays of kitty litter in the freezer and lower part of the fridge. I thought I had cleaned and deodorized everything, but I forgot the garbage disposal. This was our first mobile home with a disposal unit. The smell almost knocked us over when we opened the kitchen door. After seventeen days on the road in all kinds of weather and the heat and dampness of the eastern states, the ripe odor rocketed toward us.

We finally had it parked and cleaned. Jack opened the home pneumatically, with a hand pump attached to the hitch, and some of our new neighbors pitched in to get everything back in shape. We had to clean everything inside as it was black and filthy from the trip. The park didn't allow gas appliances of any kind for safety reasons. We drove to Sears in Parole, Maryland, and bought all new appliances. That proved to be a pretty expensive move.

The Gleasmans bought an older home in Vienna, Virginia. I found the price of homes back east outrageous. The Thomases lived in a motel in Springville, Virginia, waiting to find a house to rent and their furniture to arrive.

After barely putting our house in living order, we began entertaining visitors, as our friends and relatives in New York, Pennsylvania, and New Jersey dropped by. The Andrews, my sister Mary and her husband Charlie, were the first to come. We still didn't have a cooler, and Mary almost developed heat stroke and climbed into a tub of cool water. We borrowed Dave and Lee's swamp cooler but it didn't work well in the humid environment.

Jack finally received notice to report for work as chief inspector on the rapid transit project. He had found a shortcut to D.C., to avoid the Beltway altogether, driving right to Pennsylvania Avenue, next to the White House, and on to L'Enfant Plaza.

On June 22, 1971, the tail end of Hurricane Agnes swept through the region, forcing evacuation of the entire park to a fire station in Edgewater, Maryland. Some folks found hotel rooms, but they filled quickly and we couldn't find a room. A firehouse isn't all that big; it was a sea of people moving around and looking for a place to sit. We were given sandwiches, coffee, and a small space, big enough for cots for the night. We didn't sleep that night, but heard the news of flooding on the Susquehanna River between Kingston and Wilkes-Barre, Pennsylvania.

Officials evacuated everyone in Kingston and surrounding towns. When two dikes broke, Jack's parents' three-story home and many others flooded up to the attic. His parents managed to drive their car to high ground and stay with relatives in Dallas, Pennsylvania, before the dike broke.

We needed permission and all kinds of shots to enter Kingston. When we arrived, we found Jack's two sisters and their husbands from New Jersey already there. We all pitched in with the clean-up. What a mess! Thick mud, rotten groceries, and meat from flooded grocery stores mixed with oil from upturned storage tanks had seeped into the house.

We could have used gas masks. The new rugs recently installed throughout the house went to the cleaners, but they couldn't remove the mud and odor. It was heartbreaking to see his mother's beautiful piano, ruined and falling apart, tossed on top of the old appliances in the front yard junk pile near the street, where busy dump trucks hauled away all the debris.

We drove back to Maryland and bought supplies as we could find absolutely nothing left and no place to buy anything in Pennsylvania.

We purchased cases of paper towels, toilet paper, soap, and cleaning supplies as well as needles, pins, safety pins, and other ordinary items. I cooked enough chili to last a few days. They were happy to eat

Although our mobile home park evacuated during Hurricane Agnes, flooding devastated the Sweeneys' home in Kingston, Pennsylvania. Jack and I helped his parents dispose of flooded debris. In the top photo, the Sweeneys are seen at left. In the photo at left and below, Jack helps move a stove. At lower left is the Gateway Gulf Shopping Center.

The flooding in Kingston damaged houses, toppled underground fuel tanks, and tipped over a building in a playground on Landon Avenue, below.

homemade hot food. I think the Federal Emergency Management Agency brought small house trailers for the flood victims, each with a clean bed, small kitchen, and dining room—just right for Jack's folks. They stayed in it until the house was fixed, inspected, and livable. The rest of the helpers stayed in Dallas, Pennsylvania, through the end of October.

Jack was working on the A-11a contract in Bethesda, so we traveled to Pennsylvania only on weekends to help finish the house.

We arrived in Washington, D.C., at a historical time with the unfolding of the Watergate scandal, followed by the impeachment and resignation of President Richard M. Nixon and swearing-in of President Gerald Ford. Gas rationing forced drivers to line up for miles at gas stations, where each could purchase only ten gallons of fuel. The poor people at the end of the line often found the stations out of gas when their turn finally arrived. Several states—including Pennsylvania, New York, New Jersey, and Texas—rationed gas by odd and even days. Drivers whose license plates ended in an odd number could buy gas only on odd-numbered days of the month, and the opposite was true for those with plates ending in even numbers.

Jack's first job was as chief inspector on the Bethesda underground courthouse rapid transit tunnels, which consisted of four shafts.

We continued living in the Maryland Manor Mobile Estates off Route 103 from Upper Marlboro, Maryland, and enjoyed watching this beautiful mobile home "city" arise from a desolate area.

Driving on the back road to the Harwood, Maryland, post office seemed like something from Washington Irving's The Legend of Sleepy Hollow. Thick fog and a forest of tall, spindly trees shrouded the narrow, winding road, even on sunny days, while swampland lined both sides.

I could barely see the little dirt lanes that meandered through the trees. A narrow, wooden bridge spanned the road where the creek ran underneath it. Anyone with a good imagination could hear the clop of horses' hooves pounding on bridge boards with a headless horseman flying by. Although quite spooky, it was a delight to travel these roads at night when millions of lightning bugs flitted through the foliage of the swampland as far as the eye could see, reminding me of neon lighting on a Christmas tree.

Harwood, located on Route 253—the main road connecting lower Maryland to Annapolis and the Chesapeake Bay area—was just a little notch in the road with a post office and several small buildings. ...

Jack worked steadily and my career as an oil painting artist began taking off. I joined an art league in Bowie (pronounced 'booie'), Maryland, which invited me twice to hang my paintings in local banks as the Artist of the Month. That was a feather in my cap!

Wayson's Corner, located just off Highway 103 near Upper Marlboro, consisted of a huge restaurant with a large bingo hall behind, fresh fruit and vegetable stands, and several small homes and businesses. Wayson's Restaurant, known throughout the area for its seafood and steak dinners, served fabulous seafood and delicious soft-shell blue crab cakes, something we had never tried before. Really scrumptious!

We dined there often and everyone who visited just had to eat at Wayson's. Noticing the large walls barren of any pictures or decorations, I eventually approached the owners and inquired about hanging some of my paintings. The owner said if I could keep the walls full they would be happy to display them.

I worried that perhaps I had bitten off more than I could chew.

I didn't figure on the huge bingo hall behind the restaurant. Wayson's bussed patrons from areas in nearby states and Beltway towns on Friday and

Jack Sweeney stands at left beside Dick Horton as workers build twin tunnels beneath the Potamac River, joining Rock Creek, Maryland, with Rosslyn, Virginia, in 1972.

Saturday nights to play bingo and dine at the restaurant. Quite a few who won at bingo bought my paintings. Sometimes I painted two or more scenes in a day to keep the walls covered.

Several of the diners who liked my style of painting requisitioned paintings for themselves, including two building contractors. I painted scenes for the rooms of their "show homes."

The art league decided to hold street shows so Jack built several large pegboards where I could hang my paintings. They looked like giant sandwich boards. He set them up at the show sites before he left for work. Our first show took place in Parole, Maryland, close to Annapolis and a new shopping center with many stores and dining places. We set up on the sidewalks, preferably under the store awnings and out of the direct sun. It found the weather still stifling hot and muggy but I think the local folk were used to it. They could always dash into the air-conditioned businesses to cool off.

Quite a few cadets wearing dress uniforms, "lookie loos" from the U.S. Naval Academy in Annapolis, passed by during that long, hot day. Many people looked but nobody bought during the first day, yet we had the whole weekend for the showings. Jack arrived at closing time and helped me pack away my stuff in one of the stores for the night.

We set up our little sidewalk stalls for shows in Vienna, Virginia, and near Georgetown in the D.C. area, where people passed by to look and sometimes buy artwork.

Jack mentioned to his supervisor, Merl Miller, project manager on the courthouse rapid transit job, that I showed paintings in Vienna. I had never met Merl, a tall, slim, balding man with large bristling eyebrows, until he dropped by with his wife, Helen, to see the street show. After introducing themselves, they looked around a bit and Merl asked about one of my portrait paintings. I told them the story of "Baby Doe" Tabor of Leadville, Colorado, the character I remembered from Leadville and painted after envisioning what she must have looked like in that era.

Merl said, "Well, little lady, I lived and worked in Leadville when I was young and Baby Doe would shoot at me and others trespassing near her Matchless Mine property. I just have to buy this painting." They bought it on the spot and requisitioned me to do a portrait of Merl. I also did several paintings of lighthouses for them.

While doing these paintings we became good friends. Helen invited us to join the Smithsonian Institute and gave me a tour of one of the storage floors containing so many artifacts, some macabre and others oddities of all kinds.

Smithsonian Association members could view art and other displays from other countries before they

Artist exhibiting at bank

1973

The Bowie-Lanham American Association of University's Women's regular art exhibit at the Belair National Bank during the month of September is featuring the works of Sally Sweeney of Harwood, who has a varied selection of palette knife and brush paintings in oils and acrylics.

Originally from Rome, N.Y. she and her husband have traveled throughout the United States for the past 25 years, where he has been in the tunnel construction field, having a part in the construction of most of the major dams on the West Coast.

During their travels she has had the opportunity to receive art instructions from two noted West Coast artists, Olive Bristol of Seattle, Washington and Nevin Kempthorn of San Francisco and Sonora, California and has had painting exhibits in several states including a current display in The Fenmore Art Gallery in Up-State, New York.

Recently transferred back to the East Coast where Mr. Sweeney is employed on the METRO Rapid Transit Subway System in Washington, D.C., she plans on furthering her studies and capturing th beauty of this area in h paintings.

My painting of Baby Doe Tabor is visible above on the left in a display of my artwork on the streets in Maryland. At left is an article about my art display. Below is a painting of my mother when she was a child living in a Wisconsin children's home. I used coffee to paint the portrait.

Stories are on Page 3

The Metro cavern . . . Wow!

his Metro tunnel shaft is 70 feet deep, some 55 feet below the Potomac bottom and the starting point for tunnelling from the District under the river to Rosslyn in Arlington. The shaft was sunk near Key Bridge in Georgetown and the tunnel framing in the background is actually under the river.

Alex Bilanow's story, picture on Page 18

The photo above ran on the front page of *The Washington Daily News* April 14, 1972, showing the metro tunnels under construction. Jack is seen in front of the tunnel at the left. He is also seen in the photo at right, which was published in the same edition of the newspaper.

The Washington Daily News, Friday, April 14, 1972

Here is a view looking 70-feet up the shaft of the Metro tunnel construction site just above Key Bridge near Georgetown.

opened to the public, such as the Hope Diamond and the King Tut antiquities that filled an entire room.

The lavish displays filled several floors with waterfalls, palm trees, and waiters who walked around serving refreshments and hors d'oeuvres.

We had entered a culture we had only dreamed of.

When the courthouse job was finished, Merl and Helen moved to Georgia. We kept in touch with them for years. They had joined our Tunnel Stiff "family."

During those years, from 1975 to 1978, Gates & Fox Construction worked several contracts in the D.C. area, sinking shafts for the Rapid Transit Authority from headquarters near the Beltway about fifty miles north of where we were living. It was wonderful having Gates & Fox employees nearby and within visiting distance. Several of our original Tunnel Stiff "family" worked on this job. They offered tremendous help getting us accustomed to the East Coast, including us in their get-togethers and parties.

Dick Horton worked with Jack on several jobs in D.C. They had a lot in common, both of them from Pennsylvania. It didn't take long for him to become a treasured Tunnel Stiff "family member."

We also made lifetime friends at Maryland Manor, not Tunnel Stiffs, just neighbors who worked elsewhere, such as Bill and Mona Brooks, who lived down the block from us and operated a parts store in Suitland, Maryland. The first day we moved in, Bill offered his assistance. He had battled Crohn's disease since he was sixteen and had lost most of his intestines due to his many surgeries. Jack was the father figure Bill never had. He spent a lot of time at our house. During the gas shortage, Bill brought Jack a five-gallon can of gas every evening.

So many times Jack rushed Bill to the emergency room at the hospital for his pain and stayed by his side until he could bring him home.

Hazel Kidwell, a widow in her eighties who lived behind us, fell in love with our dogs. We visited often with her and another widow, Lola Pendleton, who lived across the street from us. Jack mowed their lawns and worked as their handyman when they needed something done. Hazel asked us to be the hosts for parties she held at Christmas and other times during the year. It was fun.

Several of the trees in Hazel's yard had died. At Christmas, Charlie Johnson and Jack decorated the dead branches with strings of lights during the night. It did look pretty and became the subject of the park's gossip.

We became friends with our rural mailman, Jim Norfolk, and his wife, Ida, who raised show cats. At their request I did a painting of Ida's favorite cat, Binky.

They were local farmers from Harwood who raised tobacco and vegetables of all kinds. Many the mornings I found a bushel basket on the porch containing a variety of veggies.

Jim asked us to watch when his tobacco was ready to be bound in bundles and hoisted to the high drying racks in the barn. The work, done in rhythm mode, reminded me of a ballet dance. Jim grabbed a sheaf of bound tobacco with his pointed gaff and hoisted it to the fellow near the drying rack at the ceiling, who then grabbed it with his gaff and hung it on the rack. It was very interesting to watch.

After nearly three decades traveling as Tunnel Stiffs, Jack and I began to think about retiring. We bought some acreage with our son, Pat, and his wife, Nancy, in southern Thurston County in Washington state. The land, originally with a Rochester address, now has a Tenino zip code. We were in the process of purchasing a manufactured home to put on our plot when Jack was working his final construction job on the twin tunnels under the Potomac River as the senior field inspector for the Bechtel Corporation.

In late 1979, Jack began experiencing health problems, such as shortness of breath and weakness. A Bechtel yearly health exam discovered some hearing loss, but Jack was fit for the work in the tunnels.

Our family physician said he had rales in his lungs—an abnormal crackling or rattling sound in the

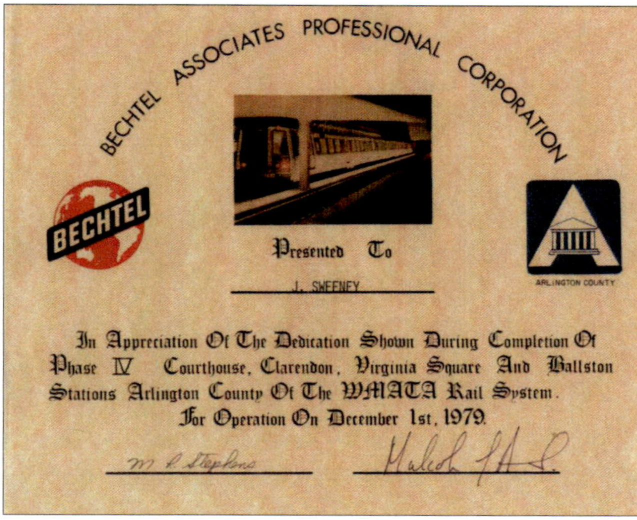

Bechtel awarded this certificate of appreciation to Jack in December 1979.

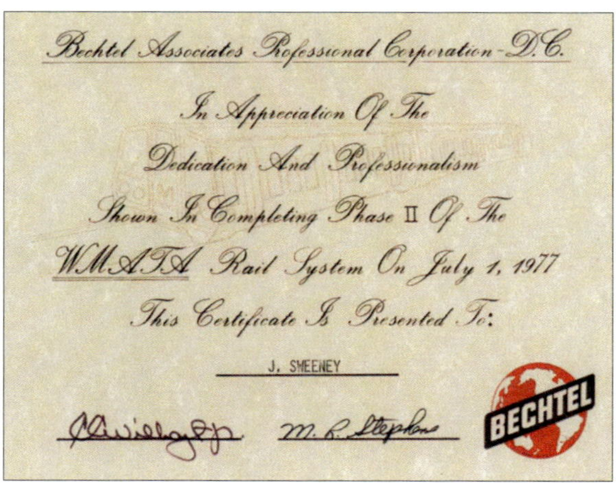

chest—and sent him to an oncologist for further tests. This was before computerized CT scans and MRIs for internal diagnostic testing. Doctors finally diagnosed Jack with cancer of the esophagus. By the time they performed his surgery, cancer had spread throughout his entire system.

Judy Gini sat with me at the hospital during his surgery. The surgeon told me Jack had only between two weeks and two months to live.

Pat flew down from Washington state. He and his dad decided to bring him to Washington to be near Pat. Mike and Judy Gini took our dogs home with them and oversaw packing up and sending the belongings we wanted to keep, including our two cars. I don't know how we could have done it without them.

Before we left D.C., we put a down payment on the manufactured home we wanted so we would have a place to live when we arrived in Washington state.

Chapter xxiii: Tenino, Washington, 1981-present

A New Beginning

I'm still living in a trailer of sorts, a manufactured home next to Pat and Nancy's house near Tenino, Washington, in southern Thurston County.

WHEN JACK WAS RELEASED from the hospital, Bill Brooks insisted on taking Jack, Pat, and me to Dulles Airport where we flew directly to Seattle, Washington. We stayed with Pat and his family until our house was ready to move into.

Shortly after we moved to Washington, on May 18, 1980, Mount St. Helens erupted, and a week later, another volcanic eruption at the mountain spewed ash throughout Western Washington, leaving the place looking like an ash-covered moonscape.

We were in our own home a little over a week when Jack was admitted to St. Peter Hospital in Olympia, Washington.

Our Gates & Fox Tunnel Stiffs from Loomis, California, flew up in a private plane to visit Jack in the hospital. It helped him so much, seeing his old friends Kirk Fox, Earl Woodward, Ed Hill, and Roy Knutson.

Norwood and Marie Driscol, who had retired and lived in Yamhill, Oregon, drove north, and Gordon and Elsie Tjerne arrived from Monroe, Washington.

Dave and Lee Thomas, who were working on a job at Stevenson, Washington, in the Columbia River Gorge separating Oregon and Washington, visited Jack several times.

Jack Sweeney, my soulmate of thirty-six years, died on June 8, 1980.

When I met Jack, I thanked God for bringing my other half into my life. We dated three months before we were married. Our wedding photo is at left. We shared love, laughter, and life together until he died June 8, 1980.

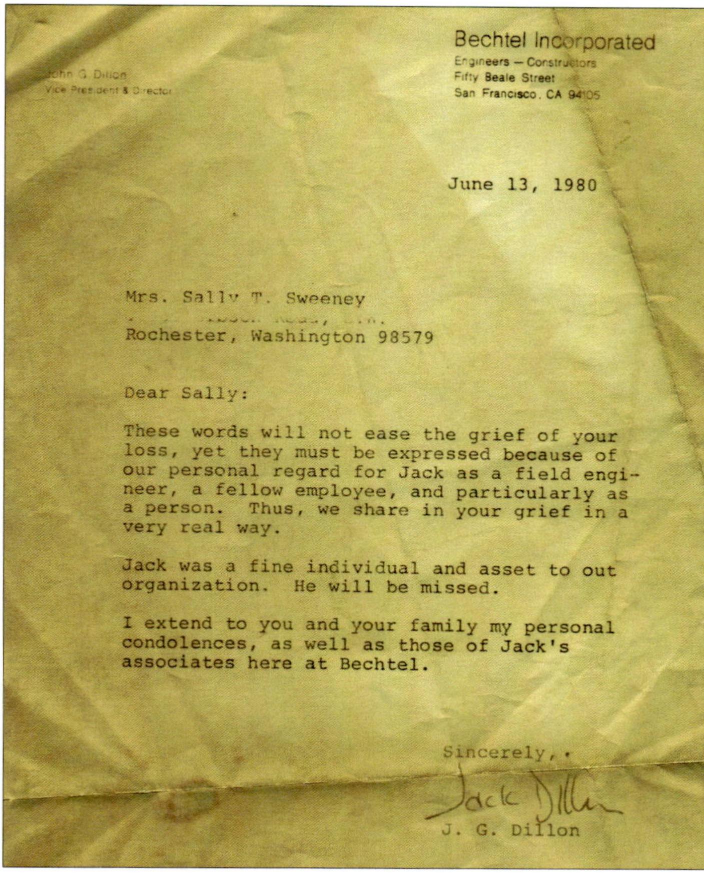

John G. Dillon
Vice President & Director

Bechtel Incorporated
Engineers – Constructors
Fifty Beale Street
San Francisco, CA 94105

June 13, 1980

Mrs. Sally T. Sweeney
[illegible]
Rochester, Washington 98579

Dear Sally:

These words will not ease the grief of your loss, yet they must be expressed because of our personal regard for Jack as a field engineer, a fellow employee, and particularly as a person. Thus, we share in your grief in a very real way.

Jack was a fine individual and asset to out organization. He will be missed.

I extend to you and your family my personal condolences, as well as those of Jack's associates here at Bechtel.

Sincerely,

Jack Dillon

J. G. Dillon

Bechtel officials sent a nice letter of condolence when Jack passed away.

The Gates & Fox crew also flew to Washington for Jack's funeral and burial. They served as his pall bearers—a true Tunnel Stiff "family" to the end.

When I finally got around to turning my house into a home, Judy Gini flew west to help me put things away and settle in. She stayed for a couple of weeks. What wonderful friends she and Mike were! They helped me more than they realized.

On November 12, 1980, Pat and Nancy adopted twin babies at their birth–Adam Patrick and Erin Rebecca Sweeney, a huge blessing in our lives and the deciding factor in the choice of where I would live. I felt very alone and lonely and toyed with the idea of moving back

We could clearly see the May 18, 1980, eruption of Mount St. Helens from our home outside Tenino.

east where my family lived. At Rochester I knew no one except my son and his family.

I made the right choice in staying here, helping with the babies and watching them grow into talented young adults.

When Erin and Adam started attending preschool five days a week, I started searching for something to do. I found a job at the senior citizens' center in Rochester, serving meals at the Lion's Club to local seniors.

I worked there for six years and loved meeting folks and making new friends. They aren't Tunnel Stiffs, but we enjoy the same camaraderie as dear friends.

Colin and Teressa Rigby invited me to go to painting classes sponsored by the Twin Cities Senior Center. I never thought I would paint again and had given away all of my equipment. I couldn't seem to return to my painting without Jack around to critique my work.

Although Elvira Dueber's birthday was a highlight of the day, there were other thing happening at last week's Rochester Senior Nutrition Easter dinner celebration. Ther was a large turnout for the traditional Easter dinner. The nutrition program for senic citizens in Rochester meets in the Lions Club building. They recently announced ne hours of Tuesdays, Wednesdays and Thursdays from 10 a.m. to 2 p.m. – Photo by Joy Ort

I worked at the Rochester Senior Center in the 1990s. I'm seen standing at back right during Elvira Dueber's 104th birthday celebration in this newspaper clipping from *The Tenino Independent/ Sun News*, DeVaul publisher, reprinted with permission.

Above is the first painting I did after losing Jack, my biggest fan and best critic. Below are portraits I painted of Adam, left, and Erin.

The first painting I did won first place at the Southwest Washington Fair, which lit a spark under me. I am still painting today.

I've made many friends at the senior center and no longer felt lonely once I settled in. Many of our Tunnel Stiff friends had retired and lived within visiting distance.

In 1981 I received an invitation from the J.D Welty and "Whitey" Phelps families to attend the fiftieth wedding anniversary celebration for Jake and Ione Semler in Oroville, California. They were friends from the first construction job we worked in Kernville and the Folsom Dam project. No way would I miss that! I caught the bus at Hedden's Pharmacy in Tenino and rode it to Oroville, where I found a room reserved for me in a motel.

I am seen with my friend, Evelyn Counts, in May 1992, outside Wolf Haven near my home in Tenino.

When I arrived I had never felt so alone and lonesome in my life. Jack and I had always been together; this aloneness threatened to overwhelm me.

Very shortly, I learned that our construction grapevine still functioned. I heard a knock at my door and there stood Miles, Pat, and Gary Moore, come to take me to dinner and visit our old friends at the Weltys. J.D. and Marg had retired and owned a small orange grove not far from Oroville.

The Big Swede and Emily Hoiland had retired in Oroville. She picked me up at the motel and drove me to their home for a visit.

Al was very sick and I felt shocked to see this huge man reduced to skin and bones. But he was still loud with his sense of humor intact. That was the last time I saw them.

The next day we attended Jake and Ione's fiftieth anniversary party and reconnected with so many of our dear old friends from across the country. I even saw George and Mary Veatch, who became our friends in Kernville in the 1940s. They had retired and lived in Lancaster, California. I remembered them all. They had quite a turnout.

Joe "Red" and Marie "Tweetie" Nickerson had retired to Castle Rock, Washington, about fifty miles south of my home. Ron and Phyllis Gleasman and family bought apple orchard property in Chelan, Washington. Dave and Lee Thomas still live in Stevenson, Washington, and Virgil and Colleen Ramsey settled in Pollock Pines, California. They would arrive in their travel trailer, park in my yard for a few days, and visit me on their trips to visit family in Colorado. Burt and Evelyn Counts from Sonora, California, did the same. They would stay for a couple of weeks at a time and we played cards and visited, so much like we had when parked next door to one another in Jimtown, California. I've also enjoyed visits from members of our family living

Pat and Nancy are holding the twins, Adam and Erin, at Christmas in the 1981 photo above, and I am holding the twins at left. Both photos are published with permission from Carroll's Photo Graphics of Centralia, Washington.

in California, New York, Pennsylvania, and friends from Ohio.

My dear friends, Lynda Orcutt from Colorado and Boots Lindquist, who now lives in Oregon, were welcome here as well.

Dot Hill and her sister Mary often stopped by for a weekend or so on their way home from visiting their family in Canada.

In 1988, Pat Moore invited me on an Alaskan cruise aboard the maiden voyage of Denmark's Princess Harmony. I flew to San Francisco near Pat's home in Pacifica, California. We boarded the beautiful ship in San Francisco and started on our twelve-day cruise.

We had a balcony off our beautiful room where we could sit and watch the scenery as we glided past it. A winding, glass stairway led from the upper deck with the shops to a huge theater showing new plays or movies every day. We enjoyed eating at several different restaurants apart from the main dining room.

We watched little dingy boats skim to the blue icebergs and collect huge hunks of ice, which were chiseled into centerpieces for the dinner tables and buffets. Real masterpieces.

We could disembark at several ports of call, look over towns we wanted to see, and buy souvenirs. I thoroughly enjoyed the old Alaskan and Yukon Klondike towns and bars, but I found Vancouver, Canada, to be rather claustrophobic with the hordes of people crowding the streets.

I especially liked the casinos with slot machines. I was pretty lucky at the slots. I found I hadn't lost my Midas touch.

Arriving at Sea-Tac Airport in Seattle, I decided to take the taxi home to Rochester, just to see what it felt like. I could learn to live like that very easily.

Thirty years have passed. I have settled down to living in one place. I do get itchy feet though, when I hear of new construction jobs and miners and their families still on the go.

I'm seldom lonely as I have made many friends in Tenino, Rochester, and surrounding areas. I'm still gathering new "families" who share my life. I can't call them Tunnel Stiffs, but they fill my life and I'm content.

Sally T. Sweeney

I am seen holding a gift I crocheted for my good friends, Dick and Kitty Thomsen, for their fiftieth anniversary September 11, 2010. Snapshots of paintings I've sold are on the wall behind me.

Friends Through the Years

Al and Emily Haviland

Bob and Freida Sheppard

Howard and Jane Van Etten

Bob and Tillie Richardson

Roy and Bertha Shoemaker

Irna and Art Thoms

Norwood and Marie Driscol

John and Marie Bartek

Dave and Lee Thomas

Rose and George Ramsey

Dawne and Kirk H. Fox

Milly and Bud Saunders

Friends Through the Years

Kim, Kenny, and Brig Trisdale

Lee and Pete Kennard

Marge and JD Welty

Gerry and Bob McLelland

George and Mary Veatch

Ione and Jake Semler

John and Jean Larch

Helen and Merl Miller

Virgil and Colleen Ramsey

Red and Tweetie Nickerson

Mike and Judy Gini

Pat Moore and son Gary

Tunnel Stiffs

Mike Gini, Norwood Driscol, Clay Paulson, Roy Knutson, James T. Brady

Red Nickerson, Al Haviland, Ed Myers, Bob Richardson, Jeanie Wilmoth

Ed Hill, Jack Sweeney, Kirk H. Fox, Ralph Winner, George Boone

Glen Jenkins, Earl M. Woodward, Dick Perley, Jim Brady, Miles Monroe

O.G. Ree, Harvey Gorham, Wayne R. Dutra, John Lindquist, Roy Shoemaker

Tunnel Stiffs

Jake Semler, Dave Thomas, Kirk Fox Jr.

Ralph Winner

Jack Sweeney, Ed Hill, Robert LeMay

Dick Simmons, Lee Kennard, Ken Wheeler

Dick Horton, Earl Woodward, Ed Hensley

Mike Gini

Norwood Driscol

Clay Paulson

Roy Knutson

James T. Brady

Red Nickerson

Al Haviland

Ed Myers

Bob Richardson

Jeanie Wilmoth

Ed Hill

Jack Sweeney

Kirk H. Fox

Ralph Winner

George Boone

Glen Jenkins

Earl M. Woodward

Dick Perley

Jim Brady

Miles Monroe

O.G. Ree

Harvey Gorham

Wayne R. Dutra

John Lindquist

Roy Shoemaker